To my mother and my sisters
who are my constant sources of support.

To my nieces
Yustina, Verity, Sakura, Saskia and Saafiya —
your natural intellectual curiosity
and excitement for learning
inspires me to teach you all that I can.

First published 2023

Siregar Publishing
www.siregarpublishing.com.au

ISBN: 978-0-6459965-1-7

Text and Illustration Copyright © 2023 Dr Lufiani Mulyadi
Illustrations by Nardin Hany

For permission requests, contact:
Dr Lufiani Mulyadi
admin@drlufianimulyadi.com.au
www.drlufianimulyadi.com.au

About the author

Dr Lufiani Mulyadi

Dr Lufi is a medical doctor who has spent almost two decades in general practice in Australia. She has also been a university lecturer, working at the Faculty of Medicine, Nursing and Health Sciences at Monash University. She has worked with the Royal Australian College of General Practitioners, reviewed publications for the Australian Journal of General Practice and has presented at various national and international medical conferences.

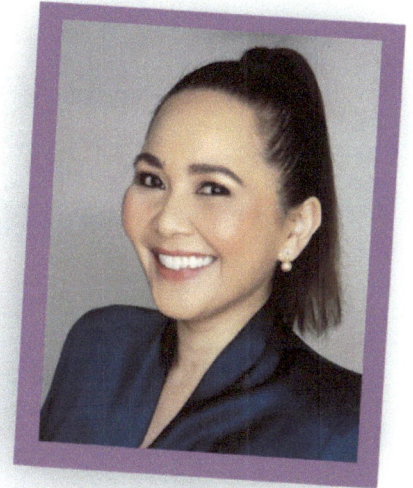

She has always been passionate about medical education and loves to teach anything medical to her patients, medical students, family, friends, colleagues and anyone who is keen to learn! She particularly enjoys teaching her inquisitive, clever and fun-loving nieces who find their aunty's amusing rhymes very entertaining!

Dr Lufi has a special interest in paediatrics which has allowed her to work with thousands of children. Through her work, she has identified a need for greater health education for children. She believes that children have the capacity to learn more about their health at a very young age, which can have a positive impact on a child's well-being and can later translate to better health as adults.

She hopes that through her books, children can explore the wonderful world of health and wellness and the exciting language of medicine. She hopes that children can learn about their body and gain confidence in understanding how their body works, inside and out, thereby improving their physical, psychological, social and emotional well-being.

A to Z of Human Anatomy

Written by Dr Lufiani Mulyadi

Illustrated by Nardin Hany

A a

ANATOMY is the structure
of the human body and each part.

A

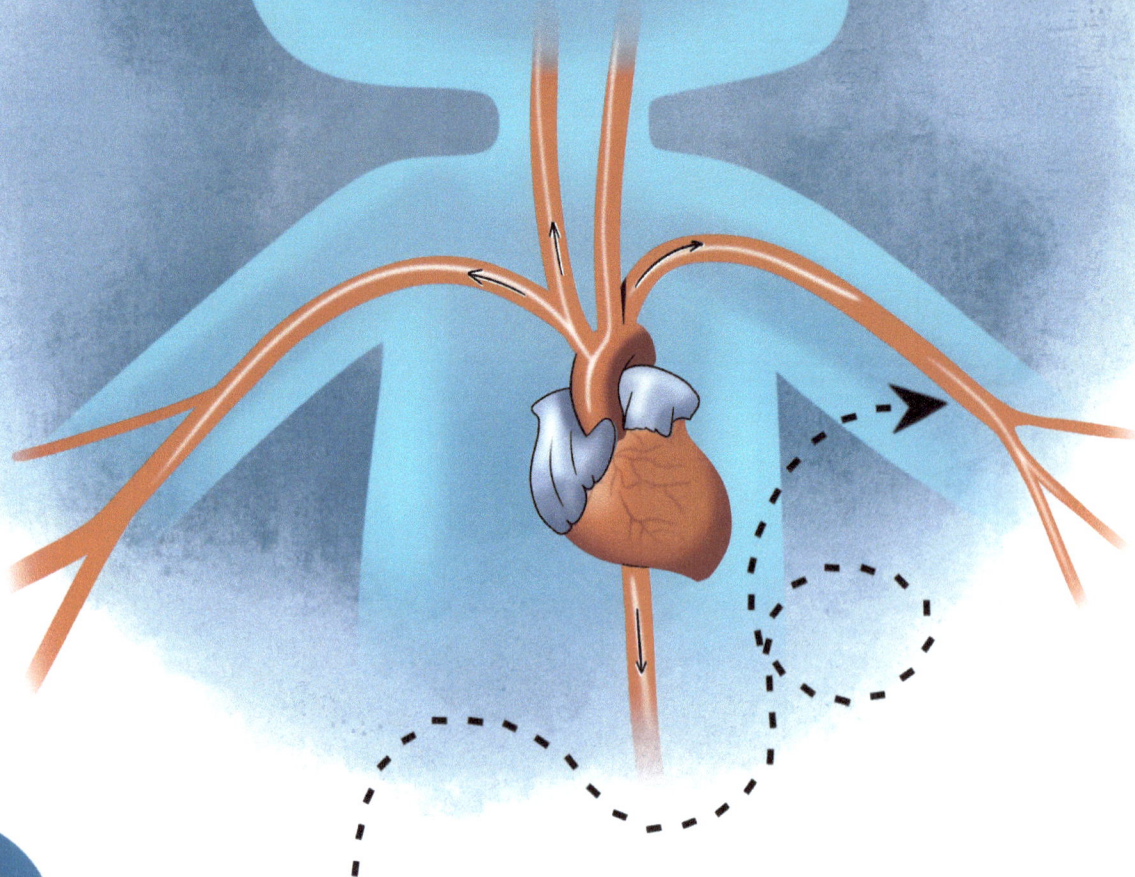

The **ARTERIES** are the vessels that transport blood away from the heart.

a

ANTERIOR is the front of everything - your face, your palms, your chest.

Anterior

Posterior

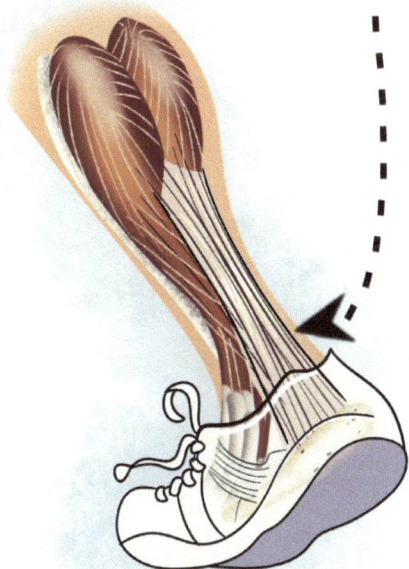

The **ACHILLES** tendon is in your ankle joint. It relaxes when you rest.

B

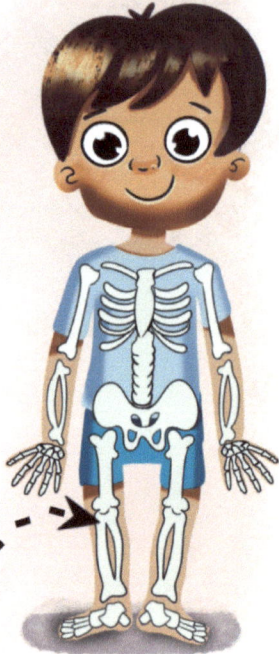

BONES make up
your skeleton,
and give you shape
even when you're still.

Your **BRAIN** controls your body
and changes your temperature
when you have a chill.

b

B

Your **BICEPS** are in your upper arms. You use them when you lift.

When your **BLADDER** fills with urine, you'll need the bathroom! You must be swift!

Toilet

b

C

The **CLAVICLE** is your collar bone. It's connected to your shoulder.

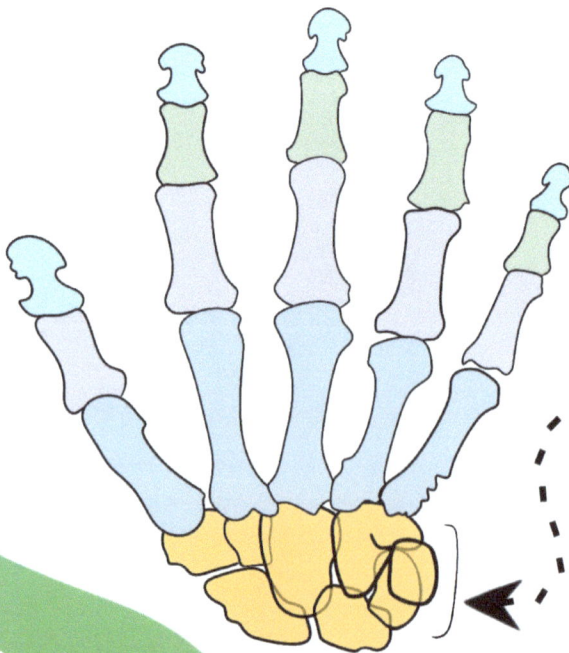

Your **CARPALS** are your wrist bones, that can creak as you get older.

C

C

Your **CORNEA** is the special layer over the front part of each eye.

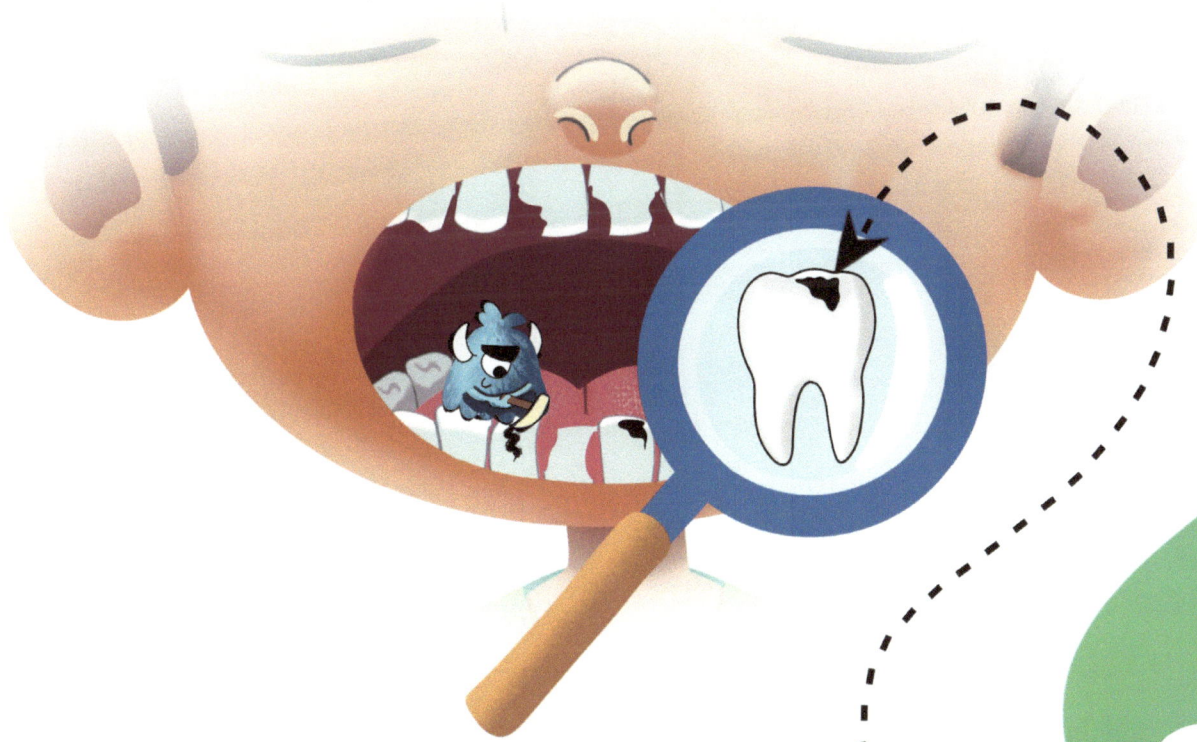

Eating candy may cause **CARIES**, so it's healthy food that you should buy.

C

D

Your **DELTOIDS** are at your shoulders. They're a muscle group of three.

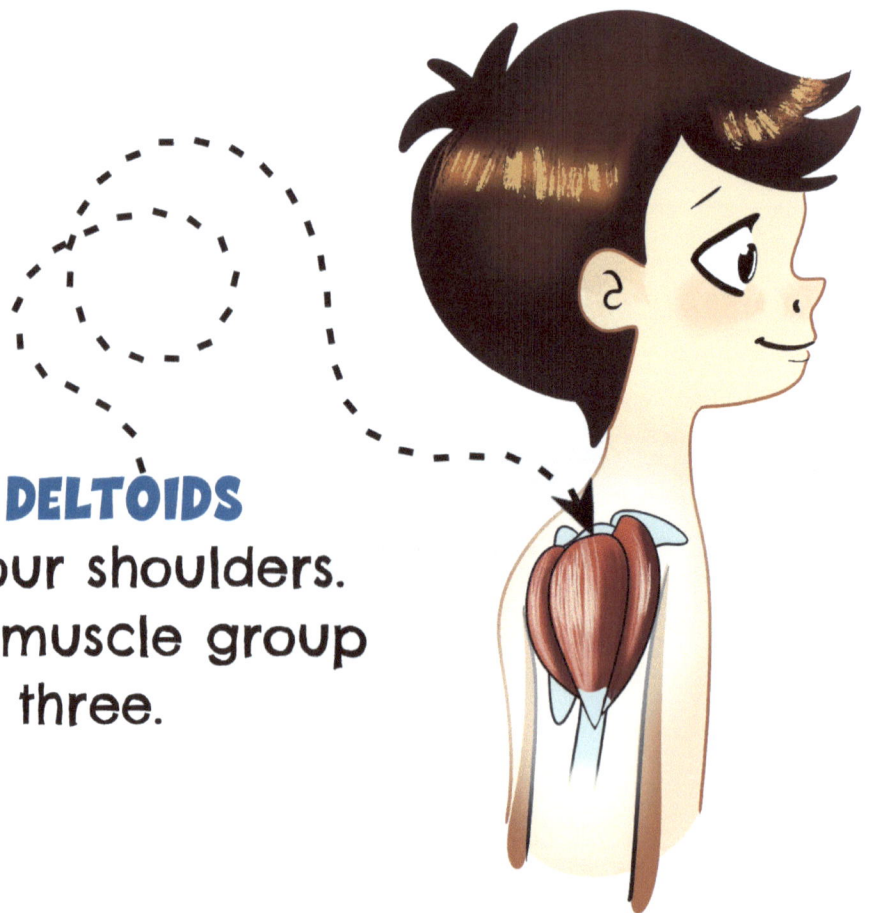

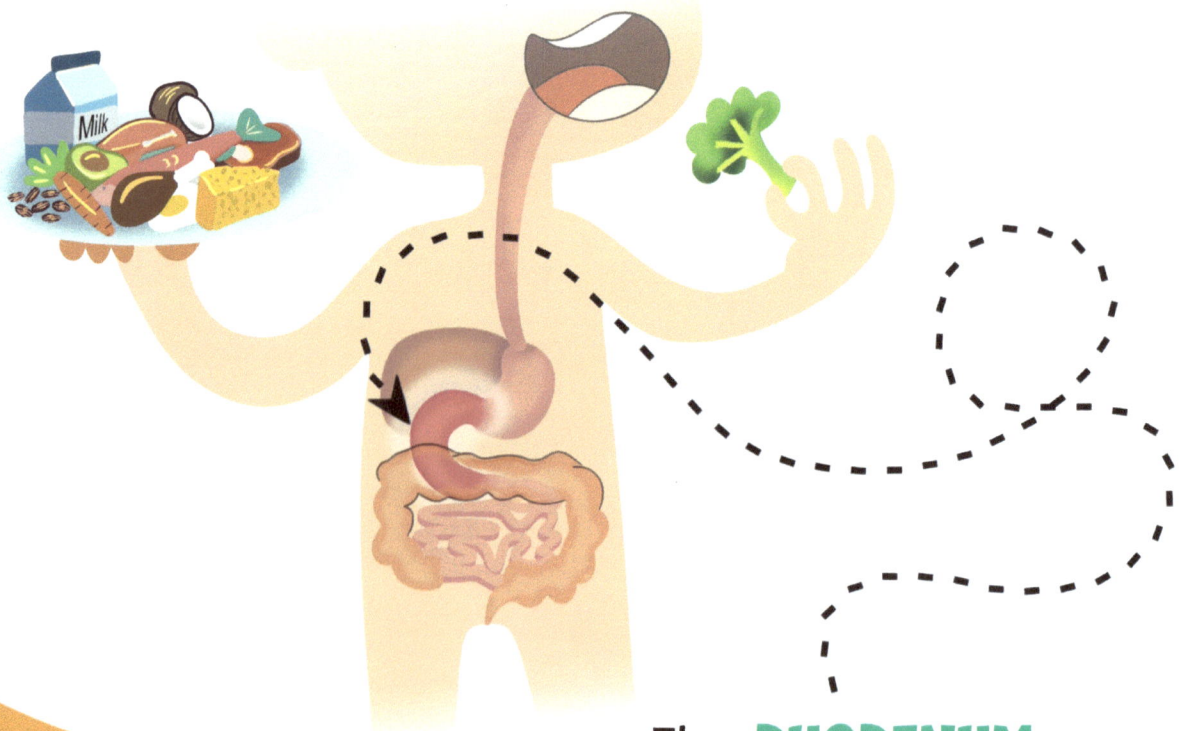

The **DUODENUM** is part of your small bowel. It absorbs iron, calcium and vitamin D.

d

The **DERMIS** is the thick, deep layer that forms part of your skin.

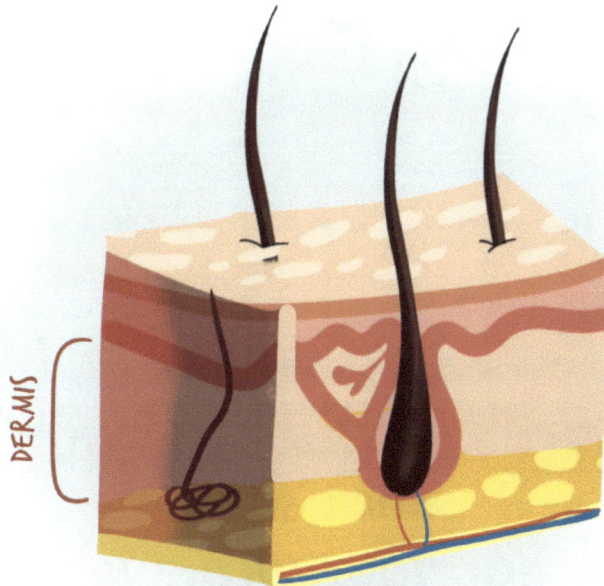

DERMIS

The **DIAPHRAGM** is a thin dome-shaped muscle that helps you to breathe out and breathe in.

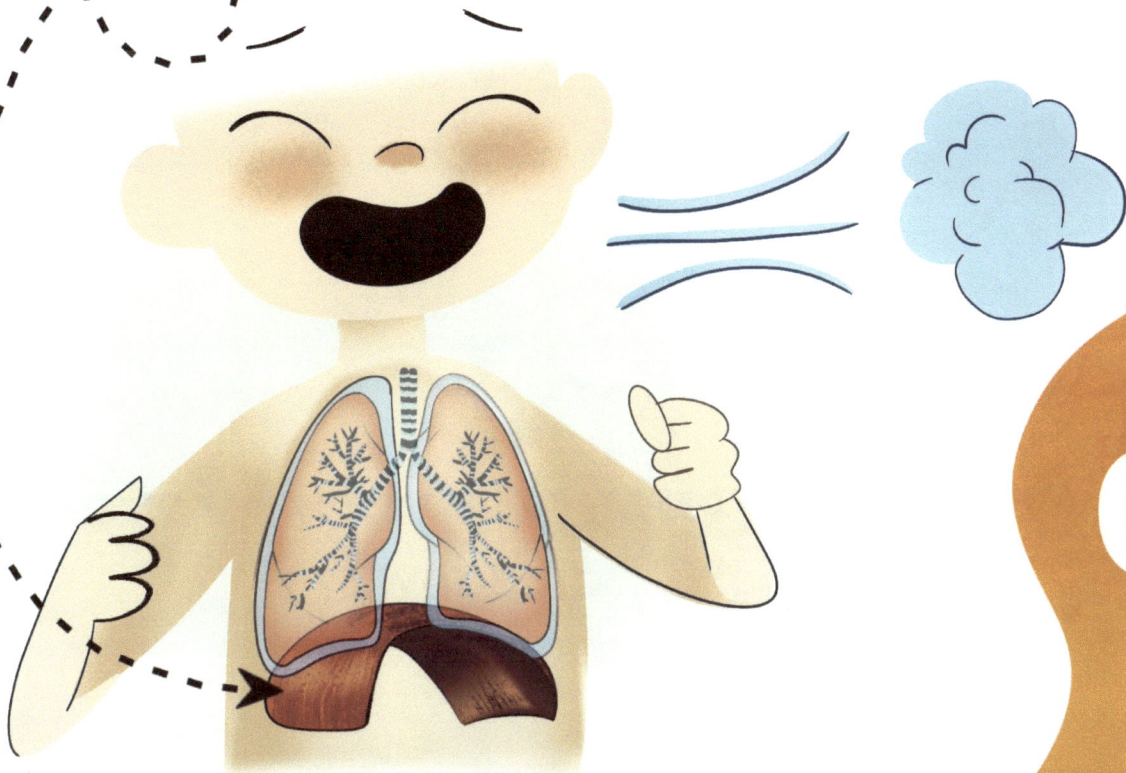

D

d

E

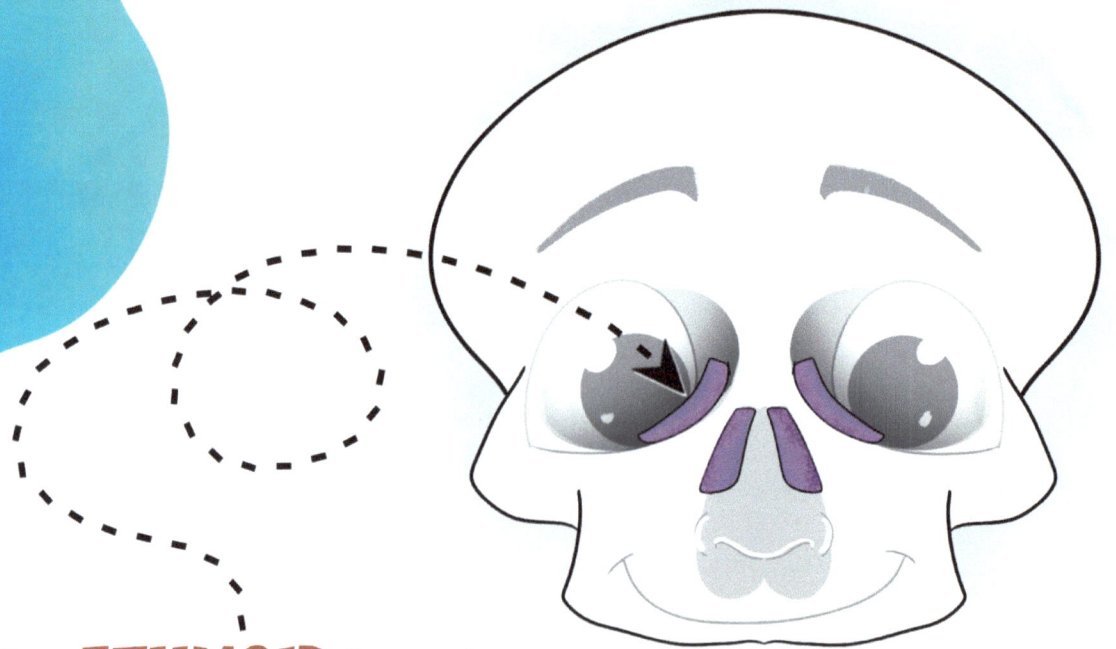

The **ETHMOID** is a bone
that sits
near your two eyes
and your nose.

The **EXTERNAL AUDITORY MEATUS,**
is the passageway
where sound goes.

e

E

The **ELBOW**
is a hinge joint,
So your arm
can extend and flex.

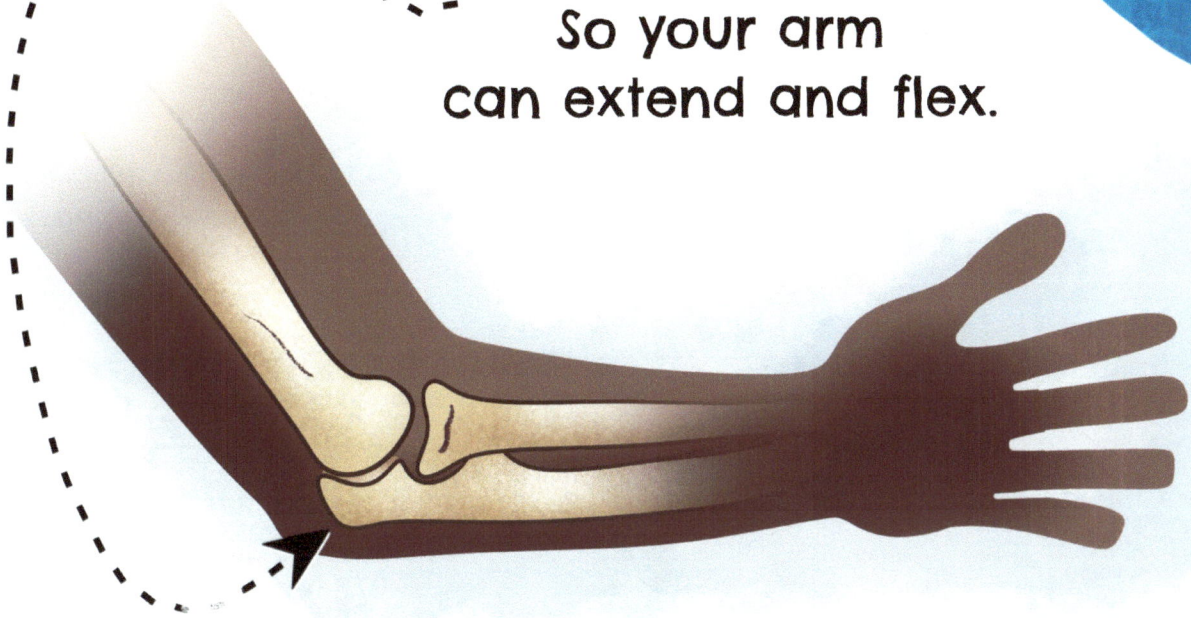

Your **EXTRA-OCULAR MUSCLES**
move your eyes
even when you're wearing specs.

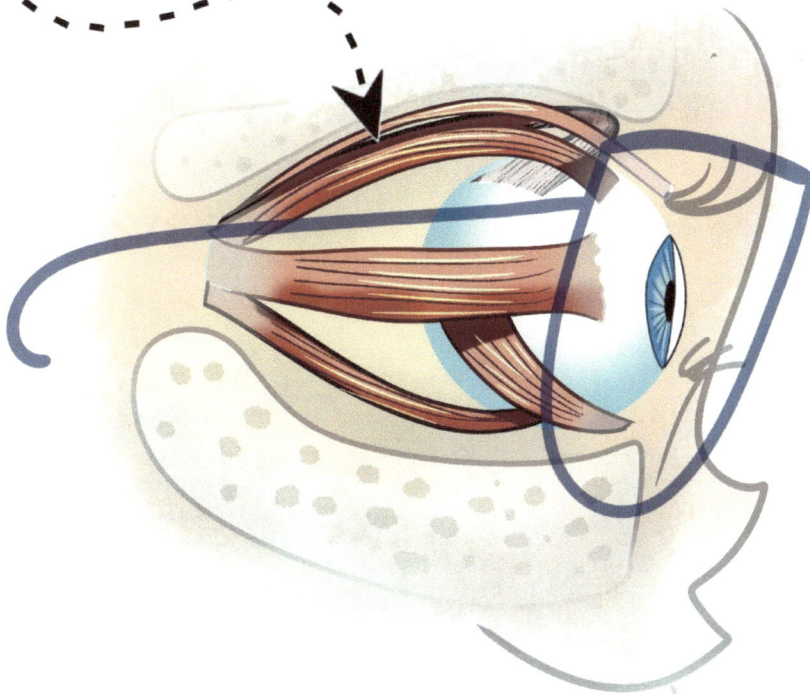

e

F

Your **FEMUR**
is your thigh bone.
It is thick
and it is strong.

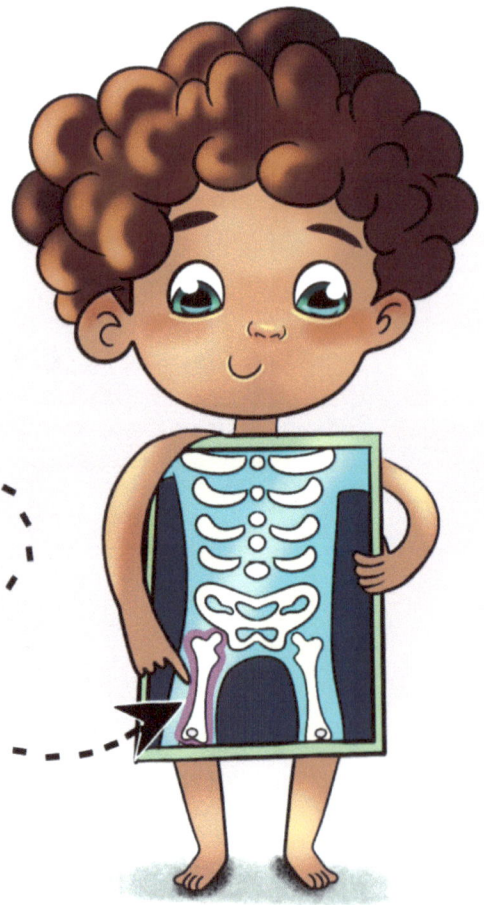

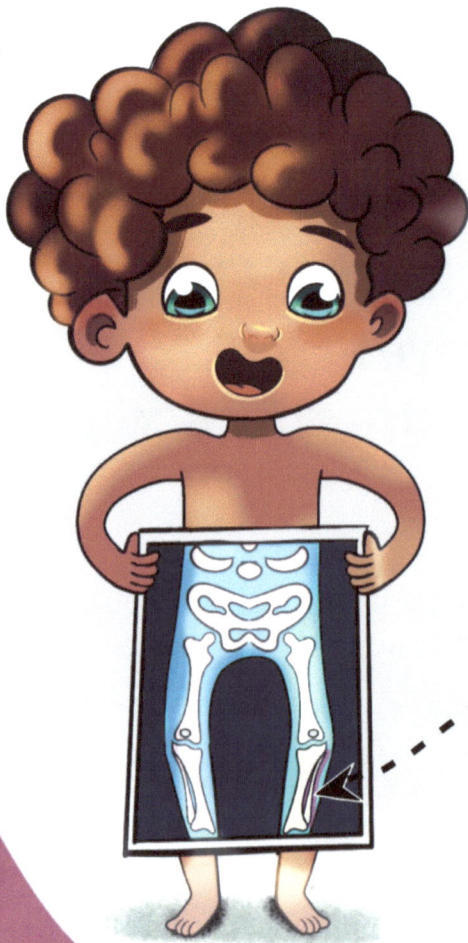

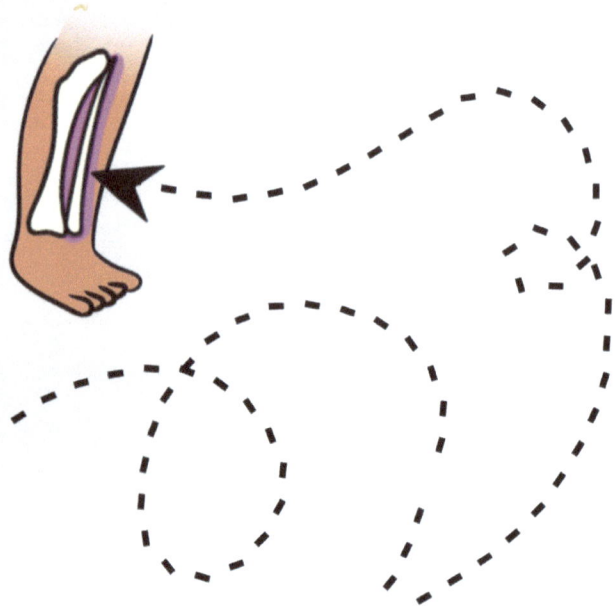

Your **FIBULA**
is on the outer leg.
It's a bone
that is thin and long.

f

F

The brain's **FRONTAL LOBE** is for planning and emotion. It really does a lot!

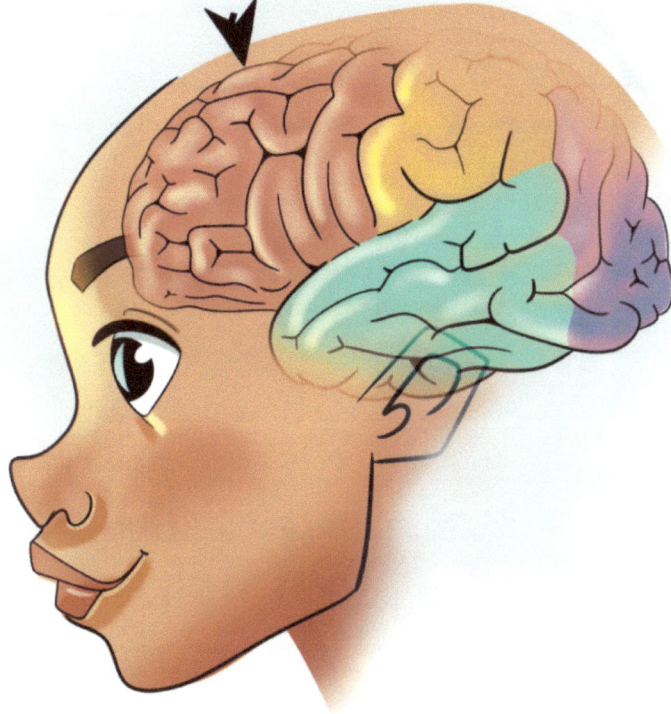

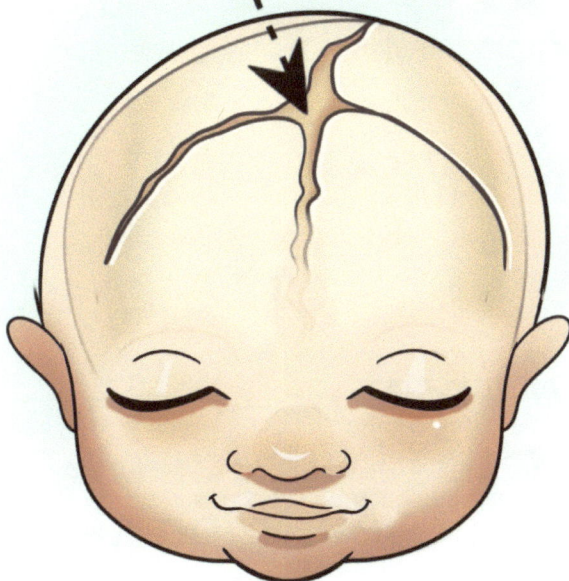

Until the skull bones fuse together, a baby's **FONTANELLE** is a very soft spot.

f

G

g

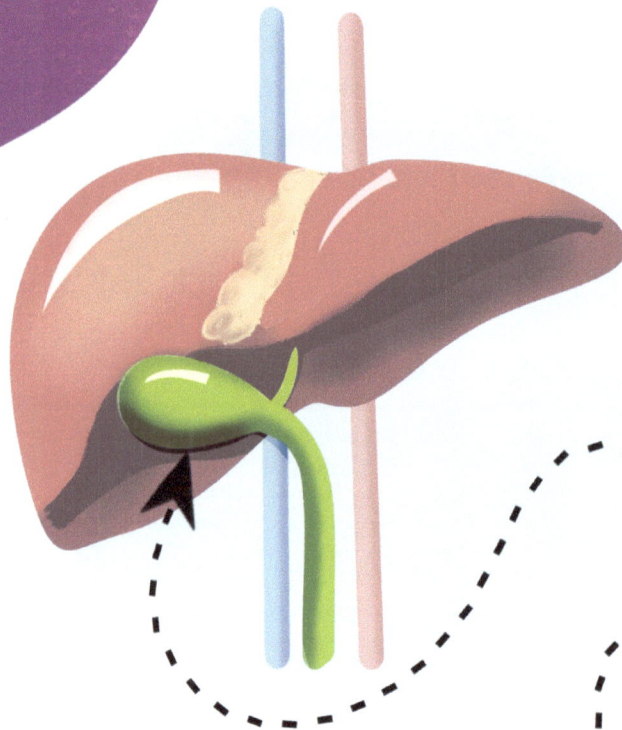

The **GALLBLADDER** stores a special liquid called bile that helps to break down fatty food.

If you upset your **GASTROINTESTINAL TRACT**, it can really affect your mood.

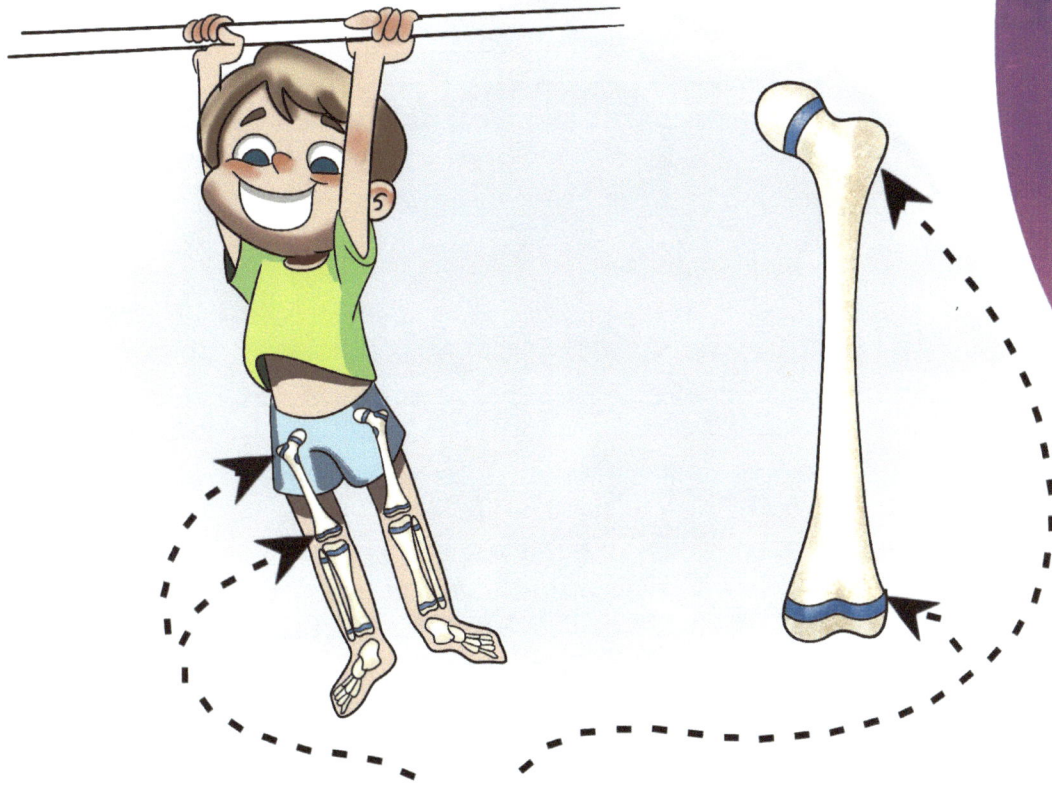

G

g

GROWTH PLATES are found
at the ends of long bones.
That's where your bones grow
and you become tall.

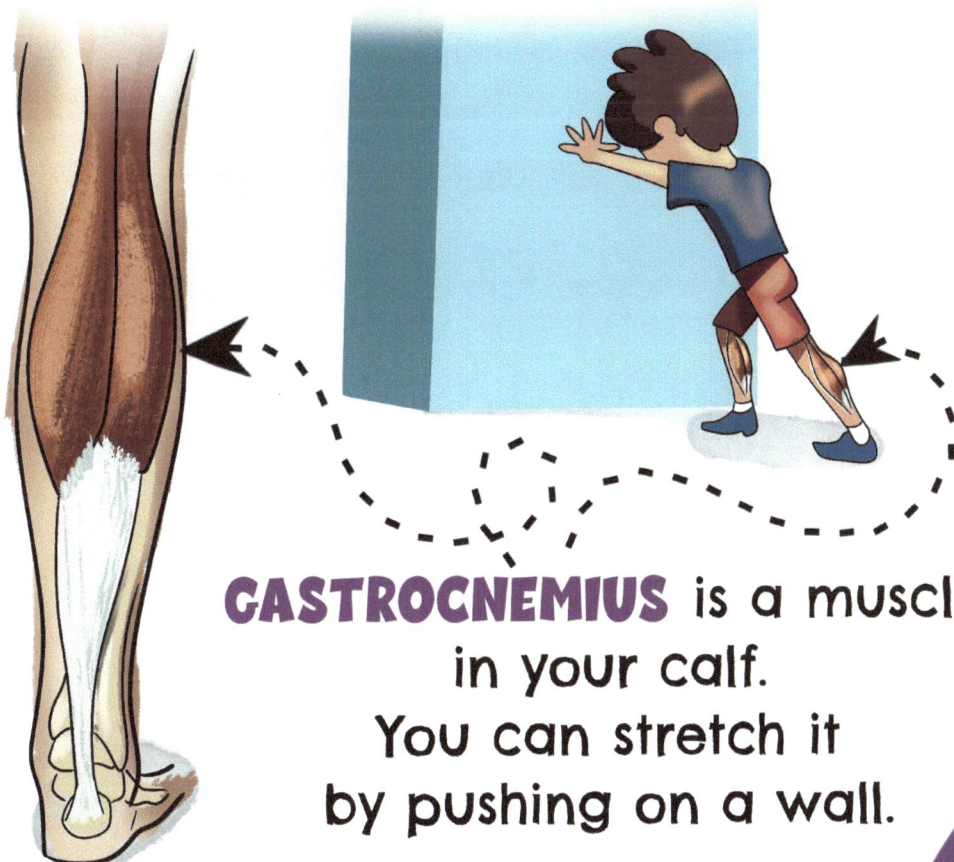

GASTROCNEMIUS is a muscle
in your calf.
You can stretch it
by pushing on a wall.

H
h

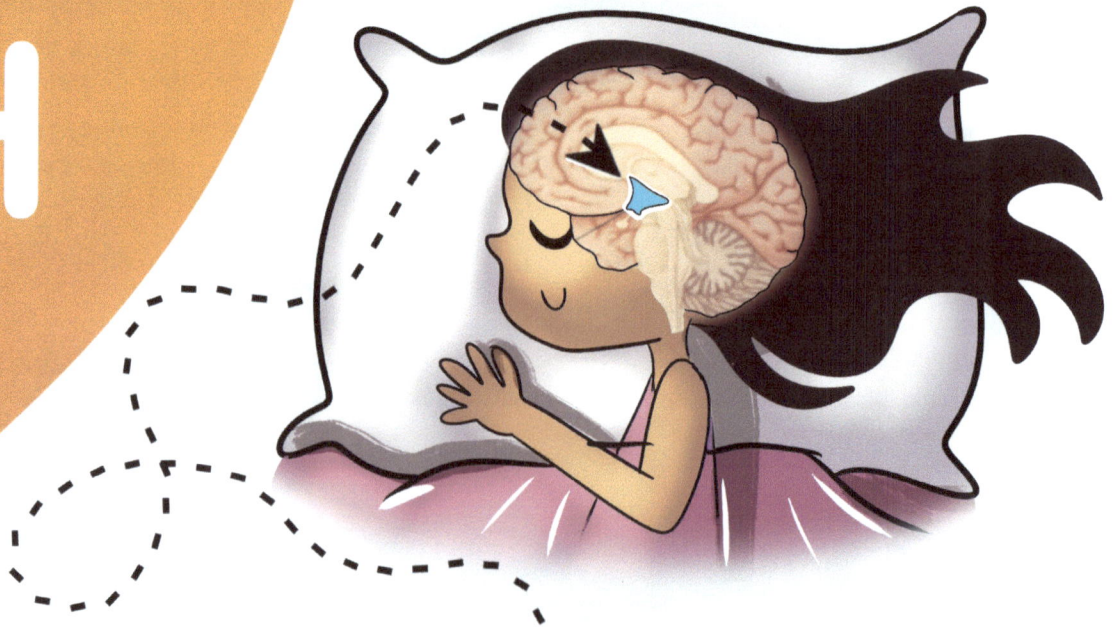

The **HYPOTHALAMUS** is located
in the brain.
It controls your temperature,
hunger and your sleep.

Your **HAMSTRINGS** are at
the back of your thighs.
They allow you to hop and to leap.

H

The **HUMERUS** forms part
of your shoulder joint,
and allows you to stretch up
and to reach.

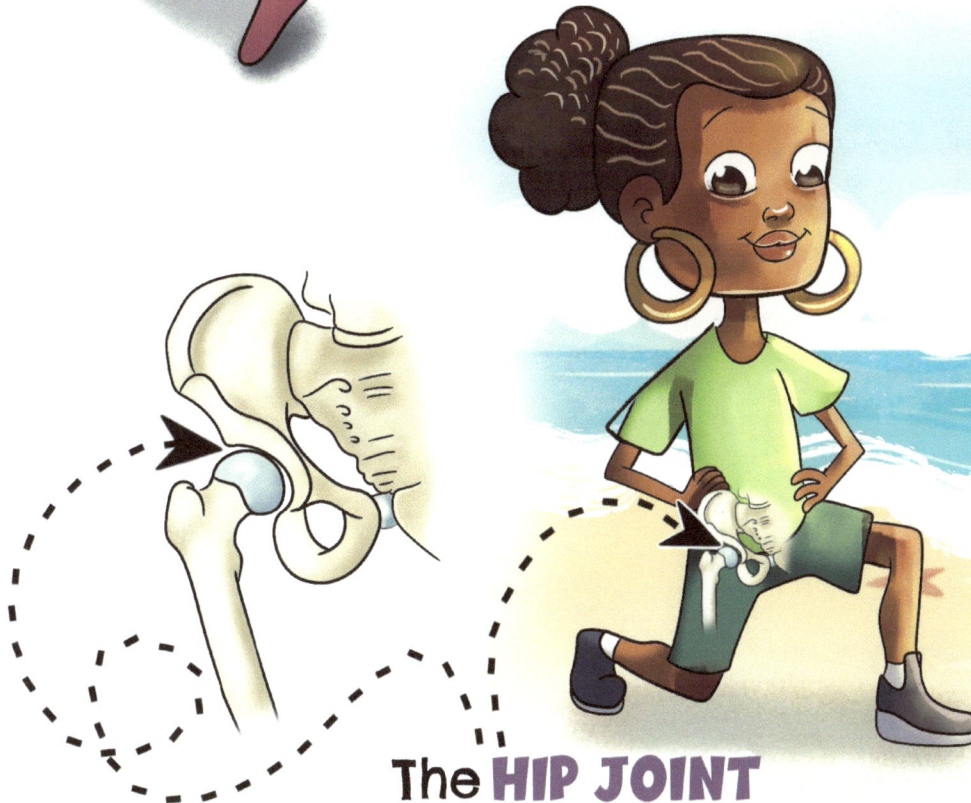

The **HIP JOINT**
is quite a big strong joint
that allows you to do lunges
on the beach.

h

I

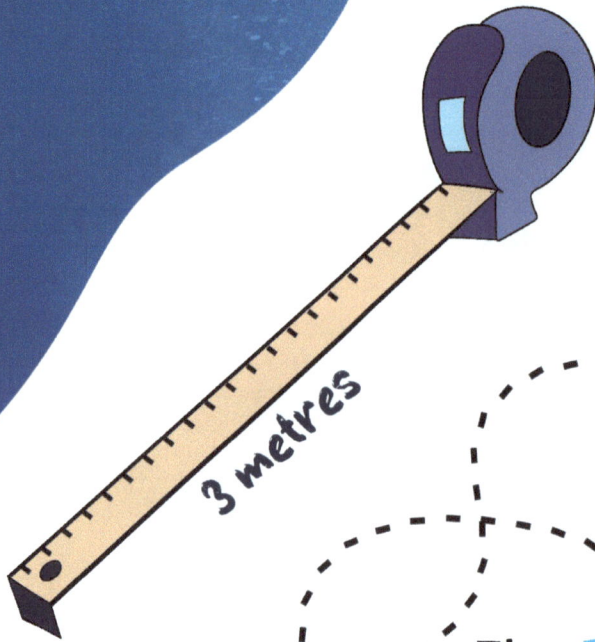

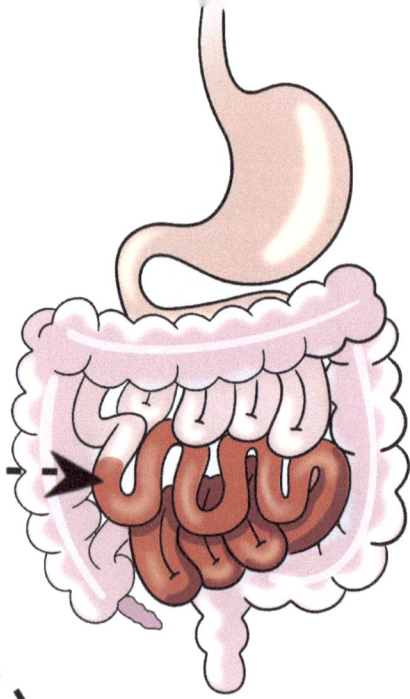

The **ILEUM** is the end part of the small bowel.
In a grown up it's three metres long!

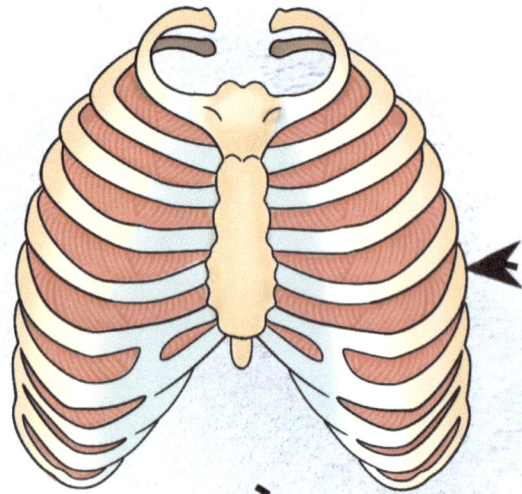

3 metres

The **INTERCOSTAL MUSCLES** between each one of your ribs help you breathe in and out during rest, exercise and when singing a song!

i

I

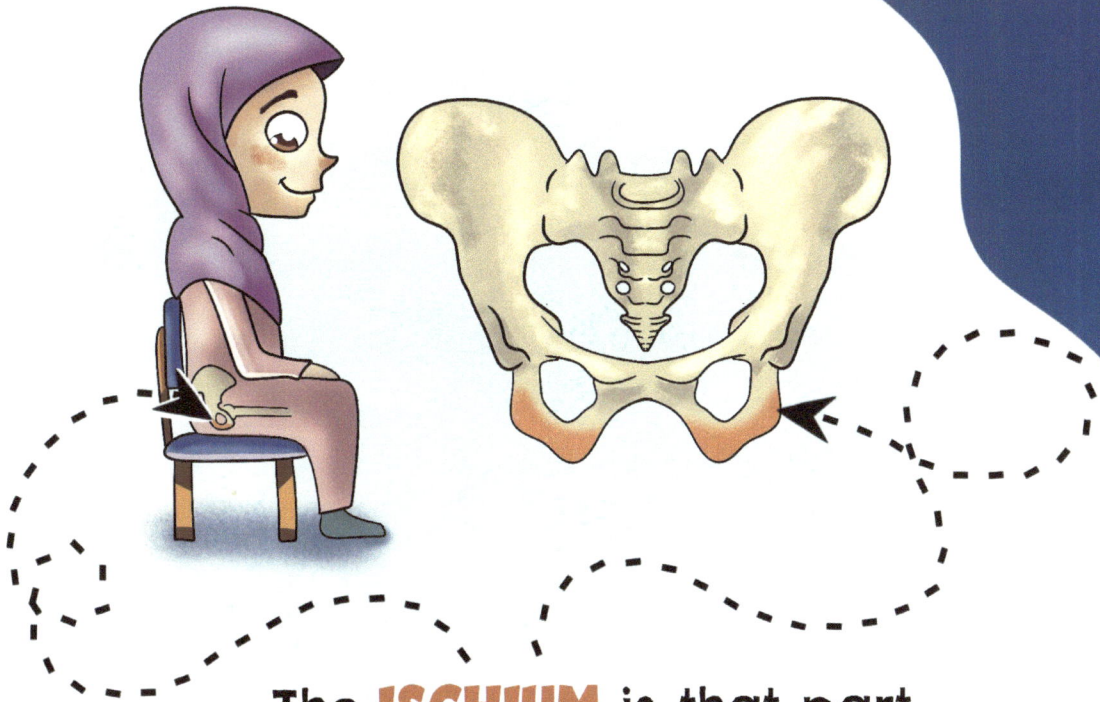

The **ISCHIUM** is that part
of the pelvic bone
on which people usually sit.

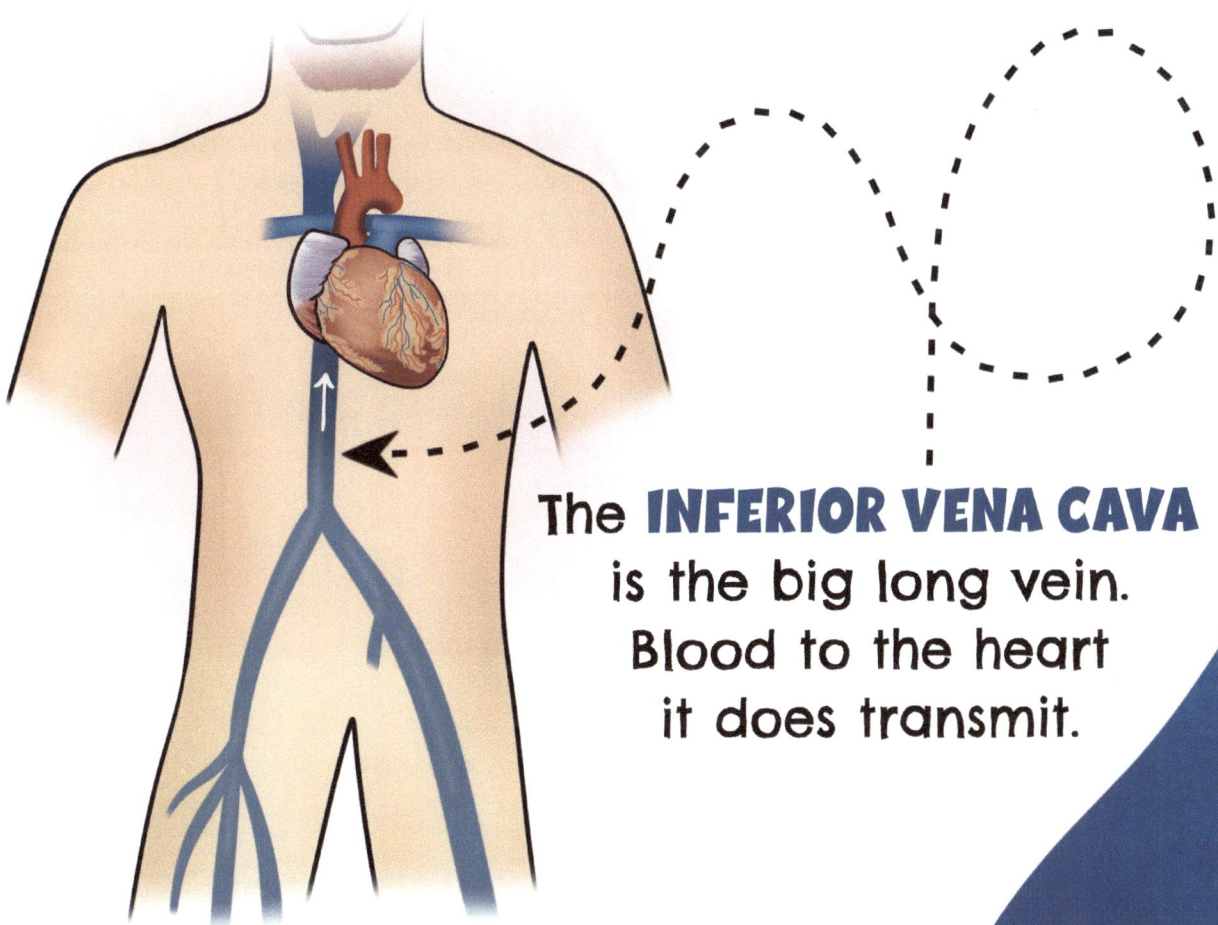

The **INFERIOR VENA CAVA**
is the big long vein.
Blood to the heart
it does transmit.

i

JOINTS bring together two or more bones to allow you to move each body part.

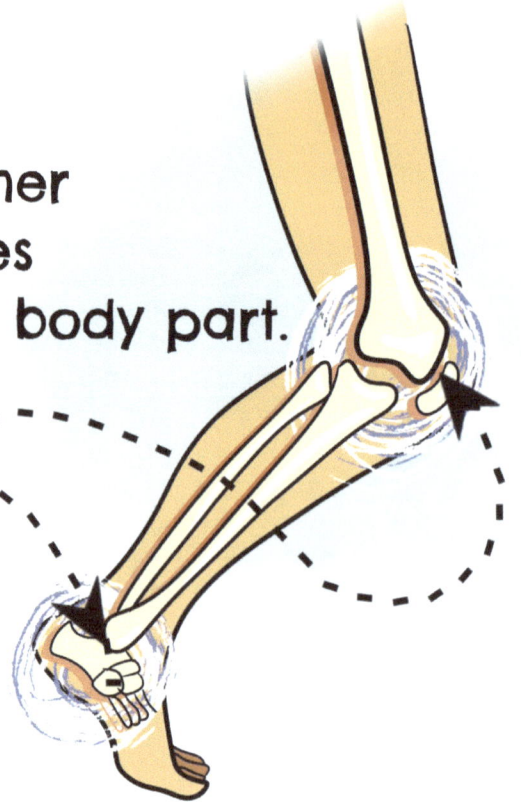

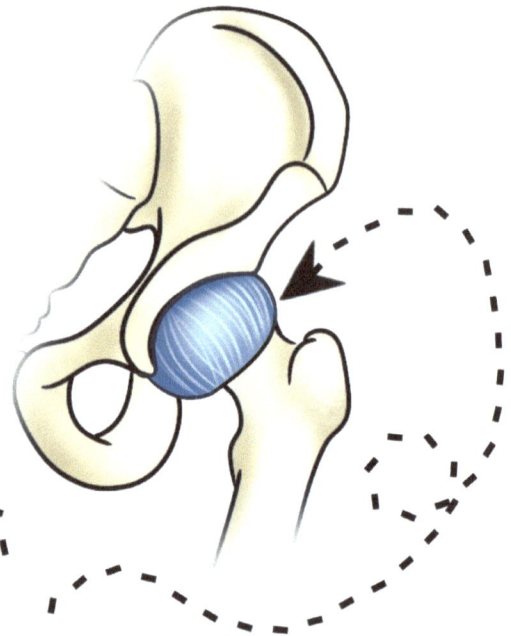

A **JOINT CAPSULE** wraps around and protects a joint and controls its movement from stop to start.

J

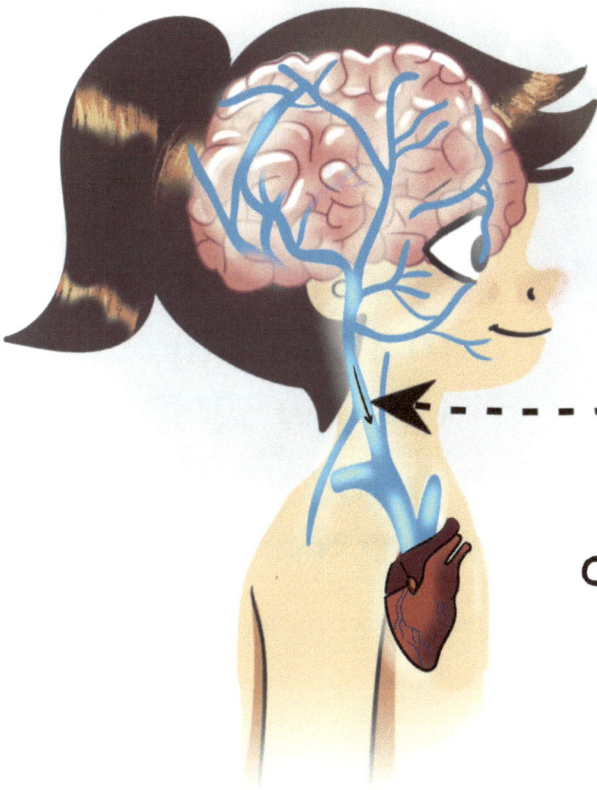

The **JUGULAR VEIN**
carries blood from the brain.
It travels down
and to the heart.

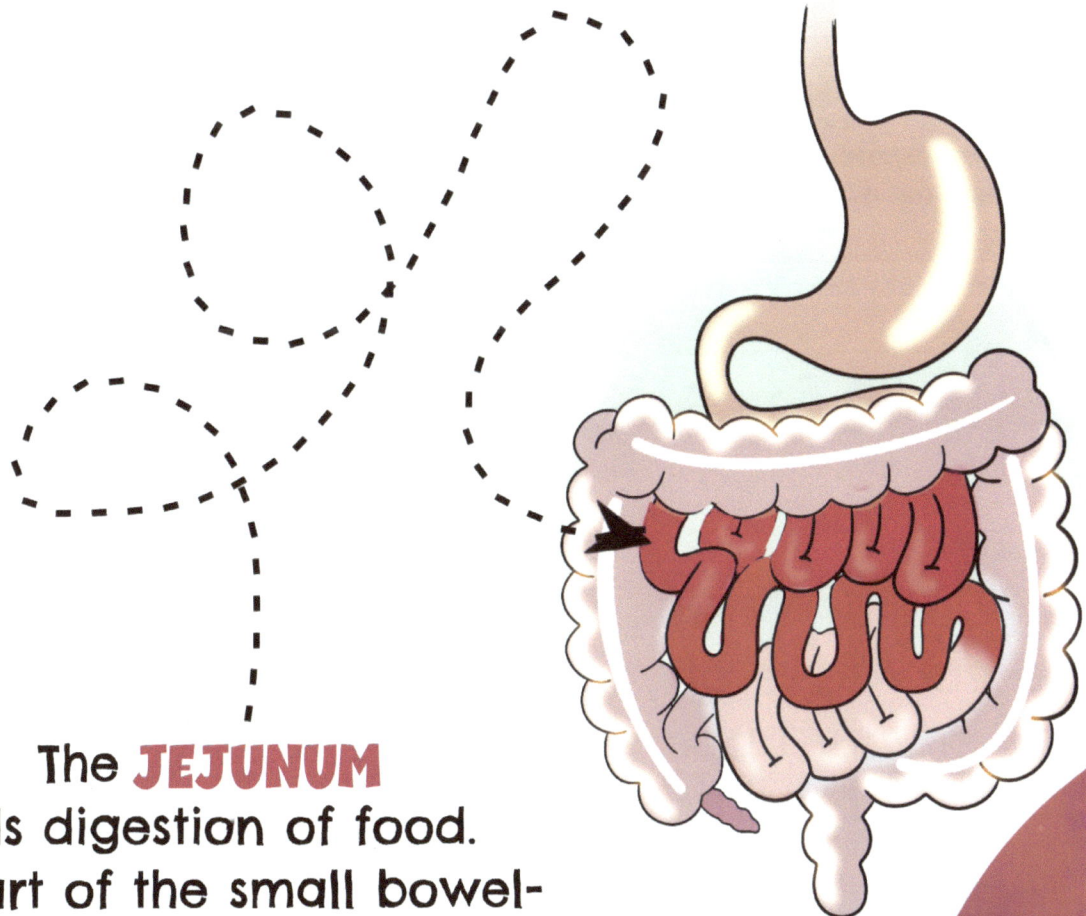

The **JEJUNUM**
aids digestion of food.
It's part of the small bowel-
the middle part.

j

K

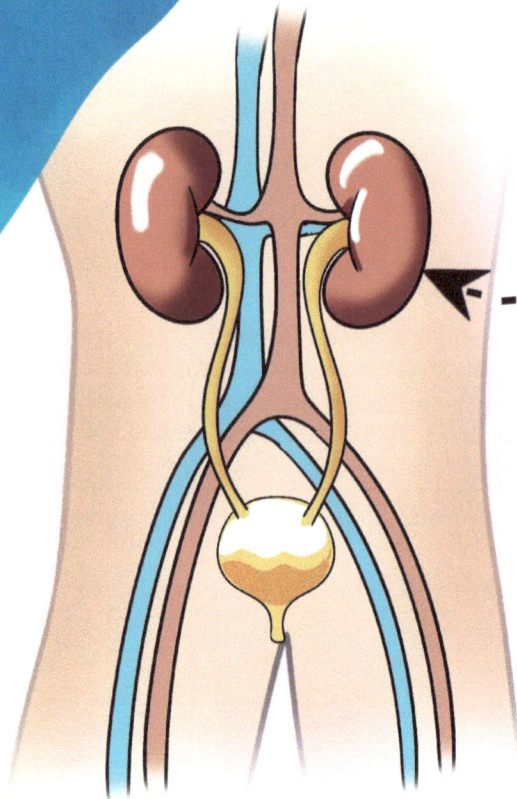

In the belly there are two **KIDNEYS**, which filter your blood as it passes through.

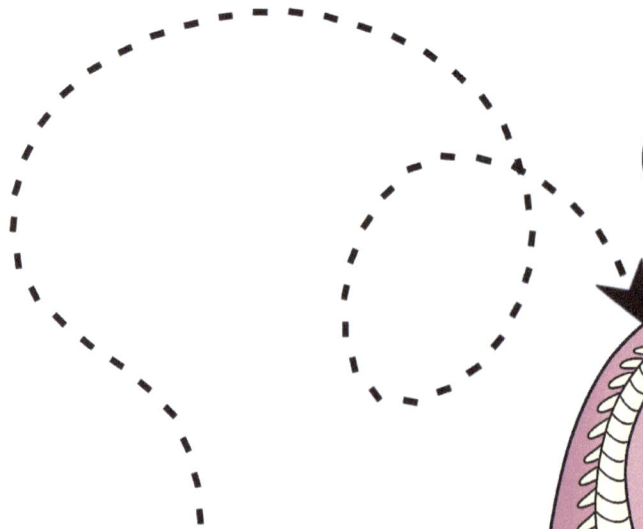

The **KYPHOSIS** is the curvature of your spine, in your upper back and sacrum too.

k

K

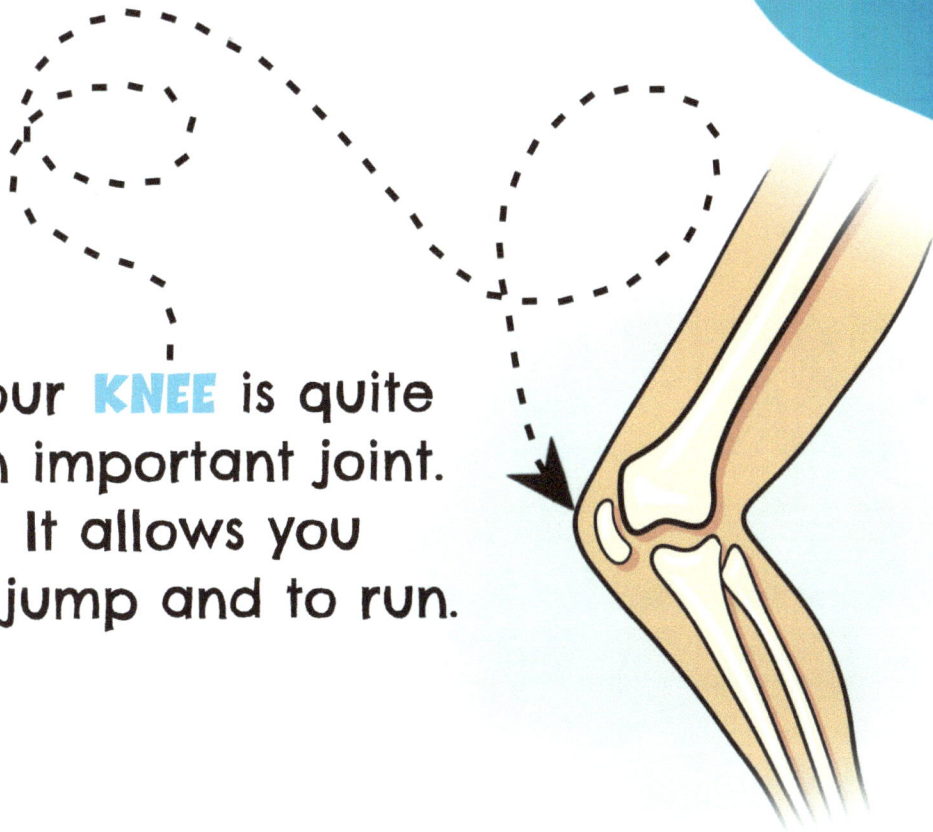

Your **KNEE** is quite
an important joint.
It allows you
to jump and to run.

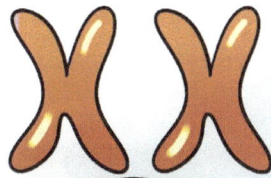

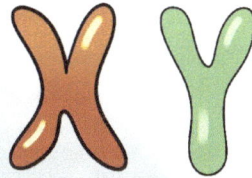

You are born with a **KARYOTYPE**
of chromosomes -
XX for a daughter,
XY if you're a son.

k

L

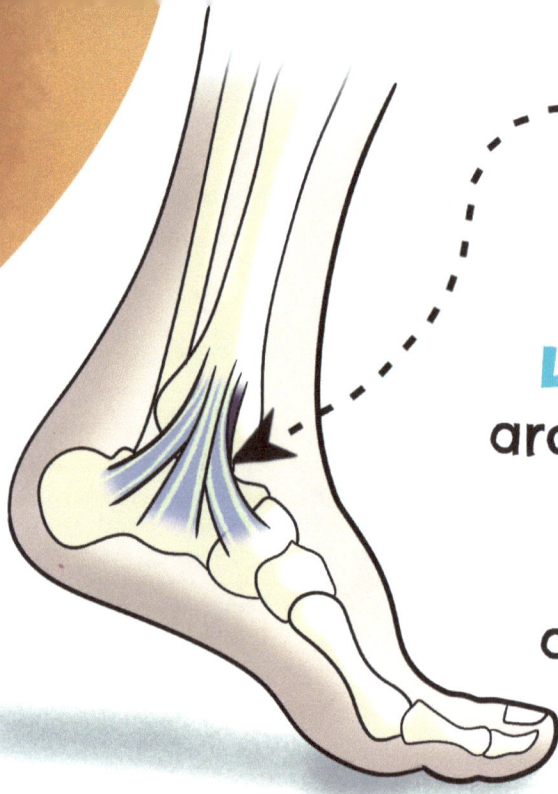

LIGAMENTS are found around all of your joints, to keep you stable when you stand and when you move.

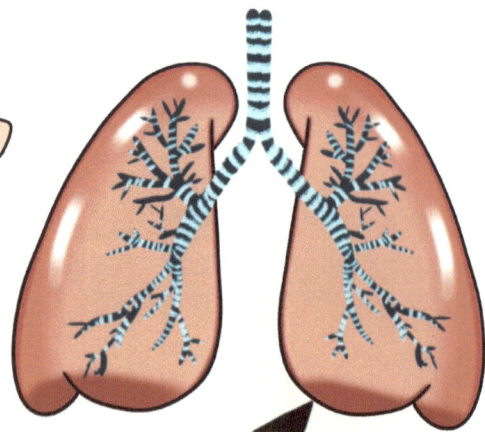

Your two **LUNGS** allow you to breathe in and out, and they work harder when you're dancing to the groove!

l

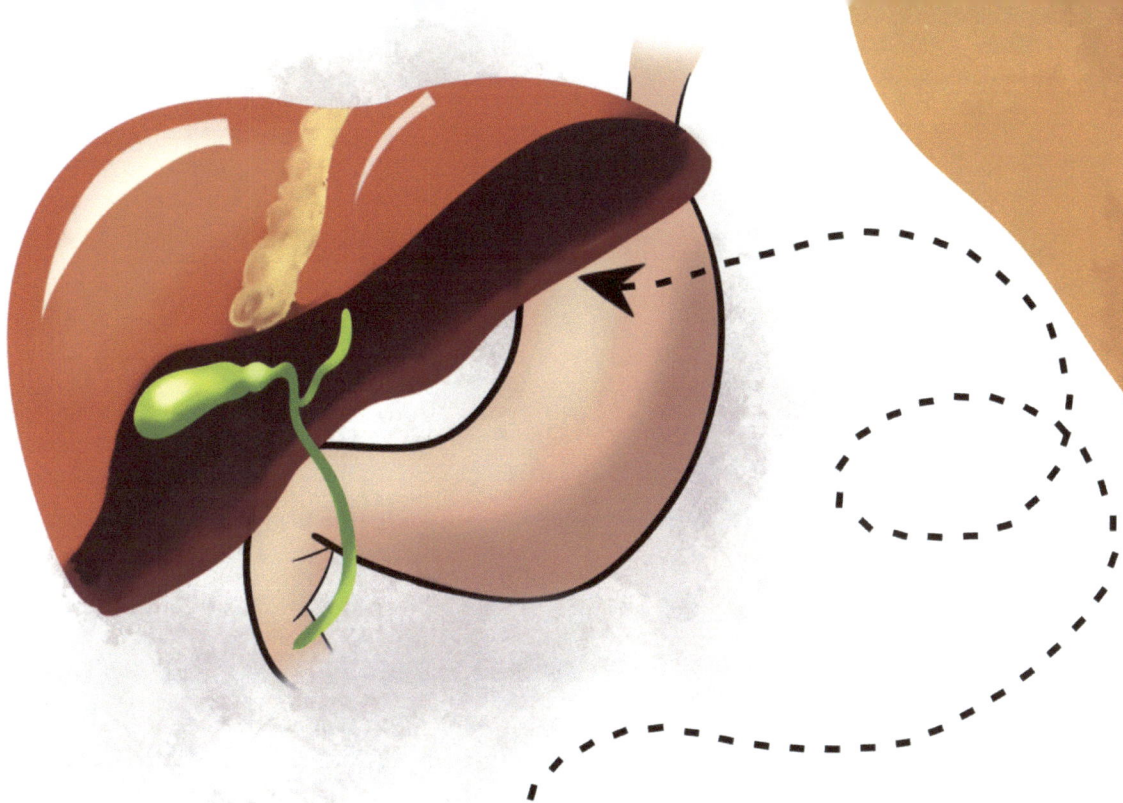

The **LIVER** is an important
organ in your belly.
It breaks down toxins
and helps your blood to clot.

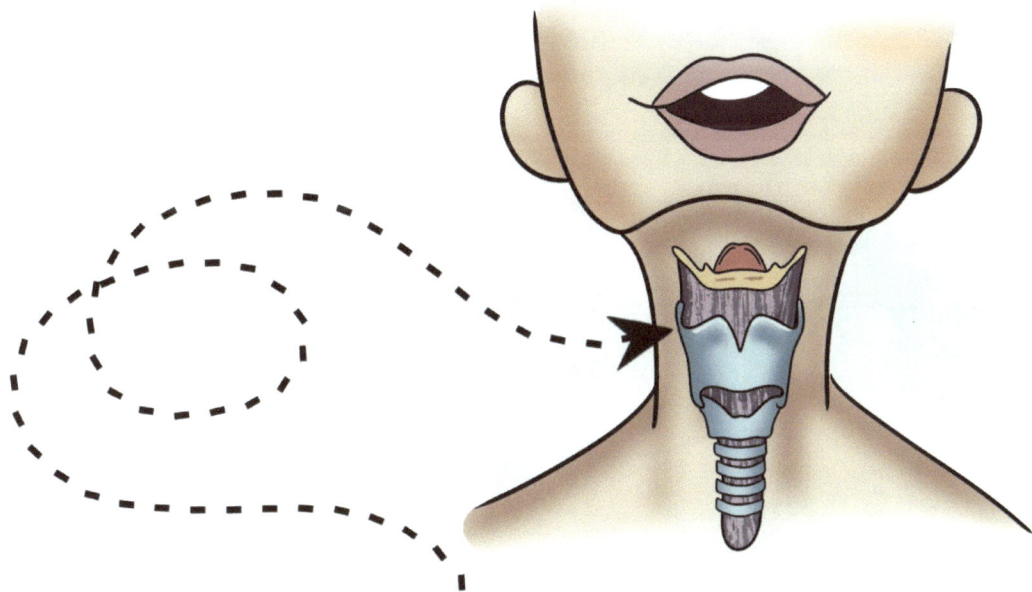

Your **LARYNX** sits
at the front of your neck.
When you talk
it moves up and down
from its spot.

M

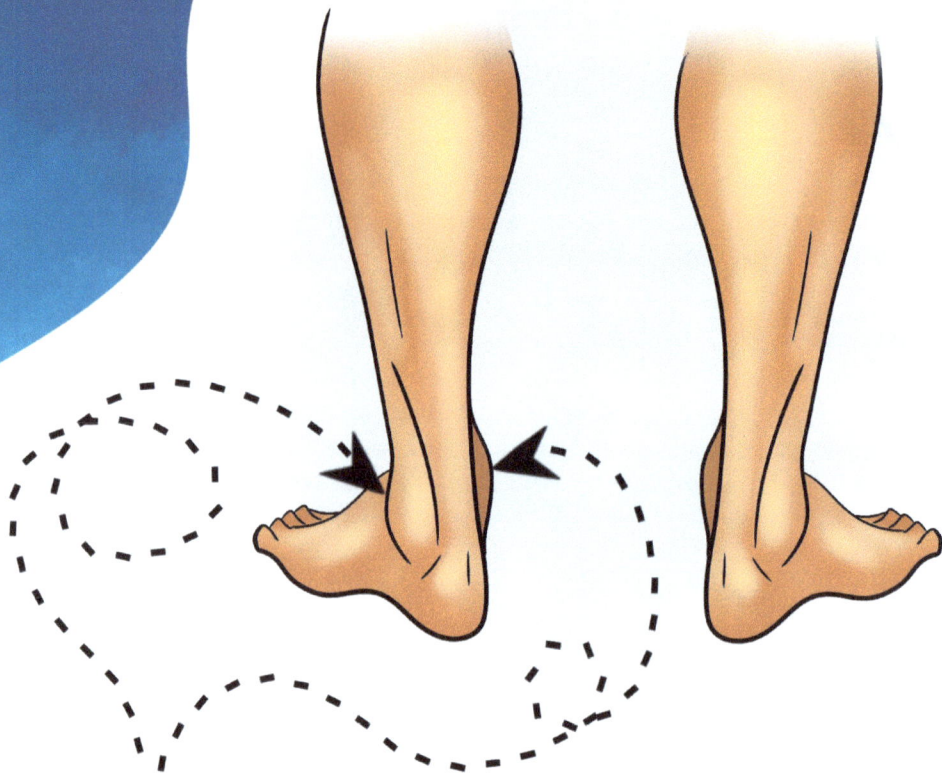

A **MALLEOLUS** is on each side of your ankle joint, so in total you should have four.

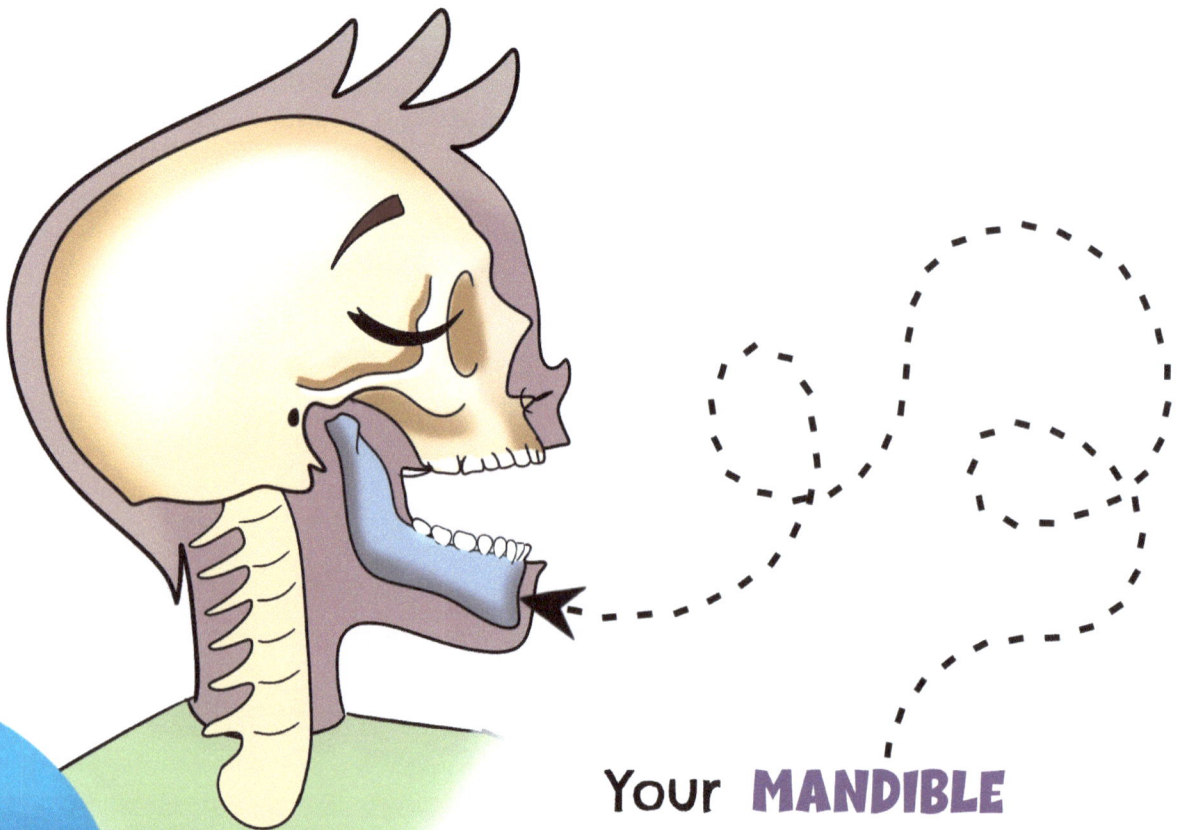

Your **MANDIBLE** opens and closes your mouth. It's the bone that forms your chin and your jaw.

m

M

The **MASSETER** muscles
in your cheeks
help you to eat
and help you to chew.

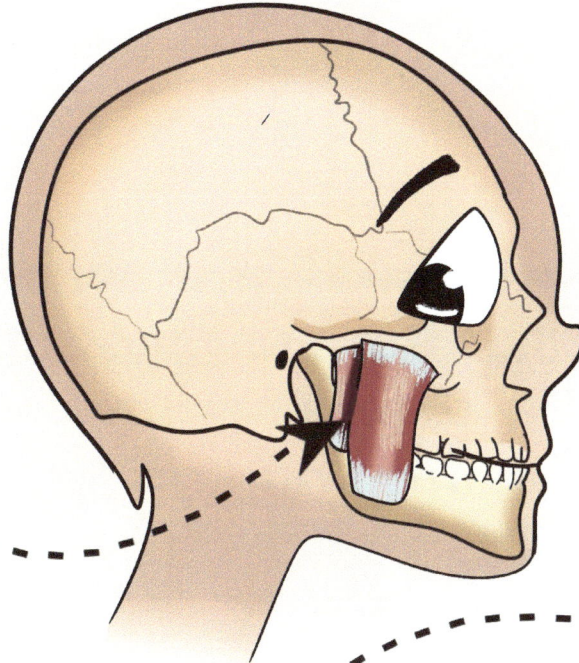

The **METACARPAL BONES**
in both of your hands
allow you to stir your special stew!

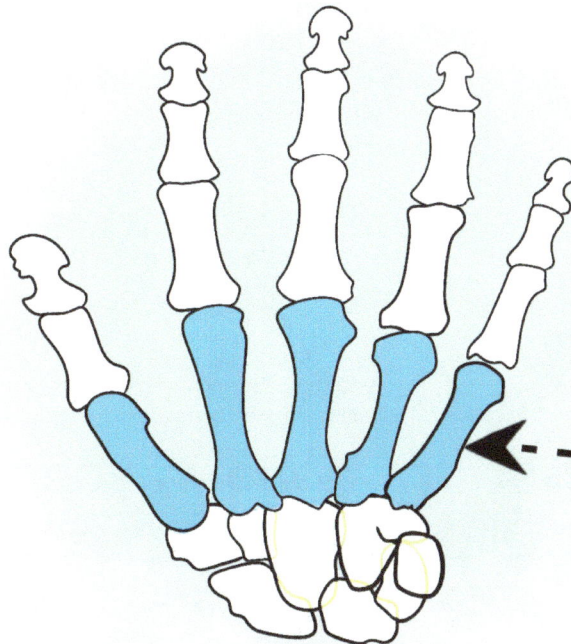

m

N

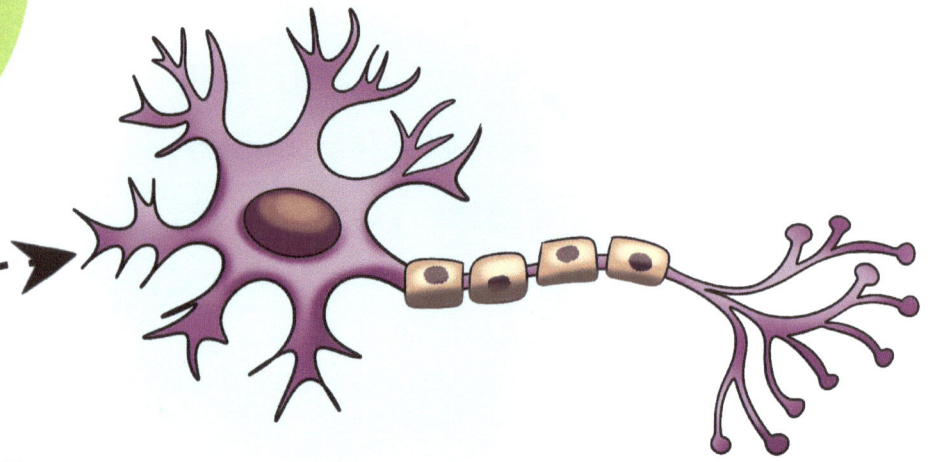

A **NEURONE** is a nerve cell
that transmits messages,
so you can move
and you can feel.

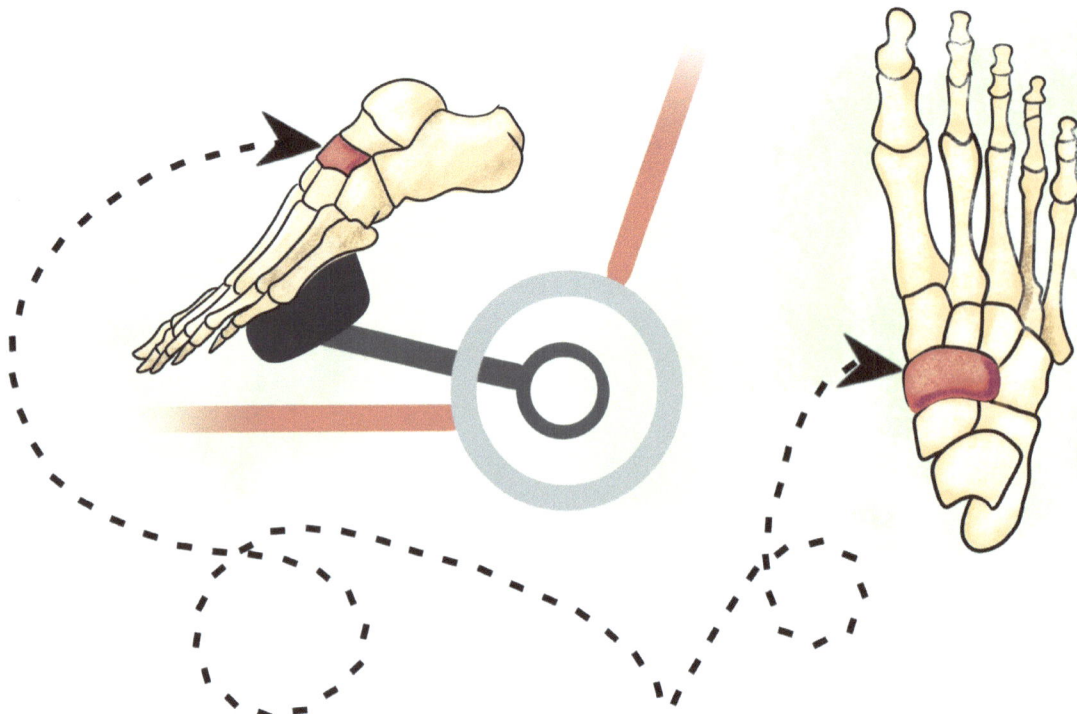

The **NAVICULAR**
is an important bone in your foot,
that allows you to pedal
with your toes and your heel.

n

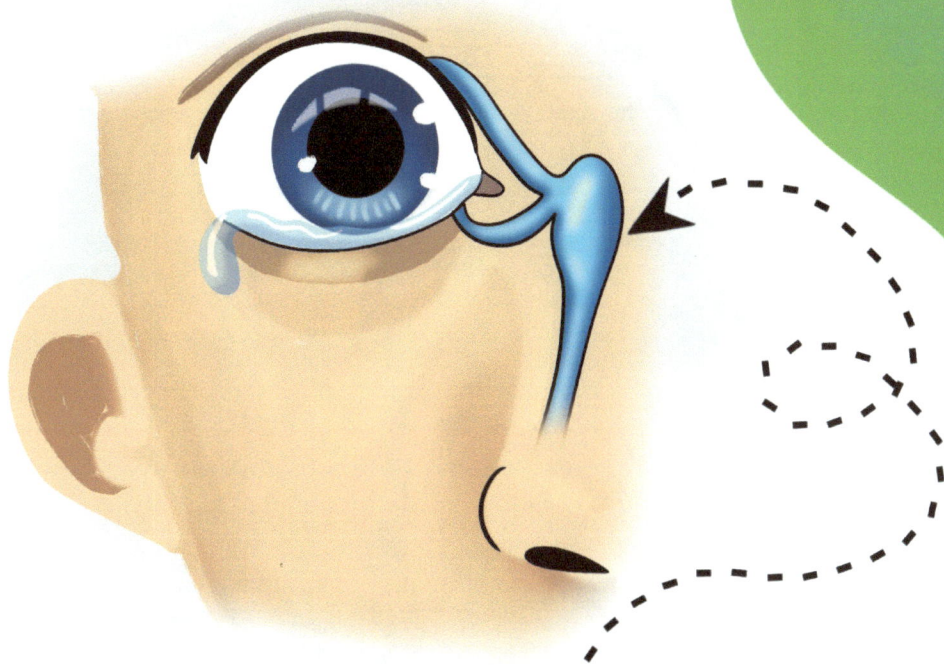

The **NASOLACRIMAL DUCT** allows tears to drain from your eyes into your nose.

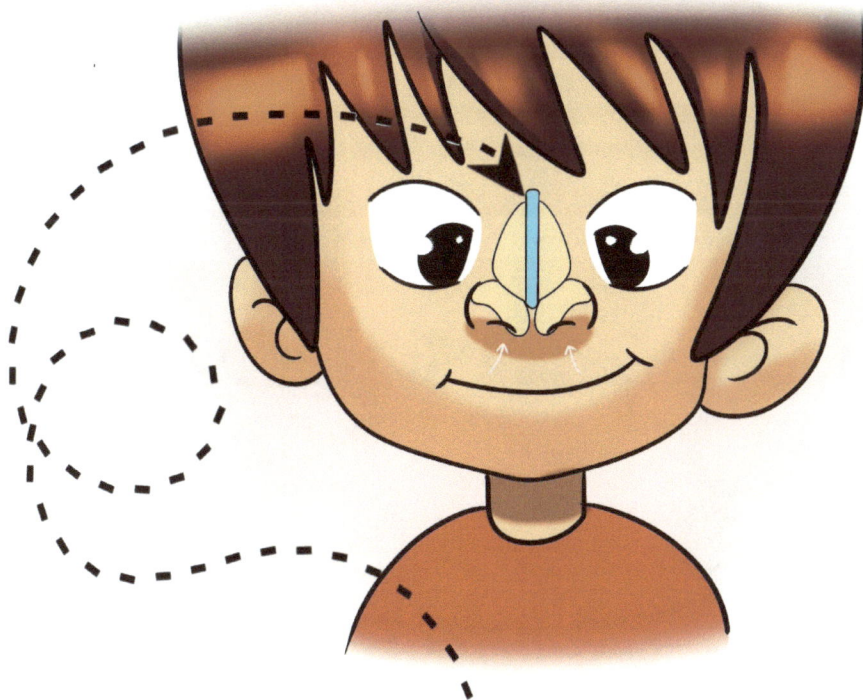

The **NASAL SEPTUM** divides your left and right nostrils. When you breathe, in and out the air flows.

N

n

O

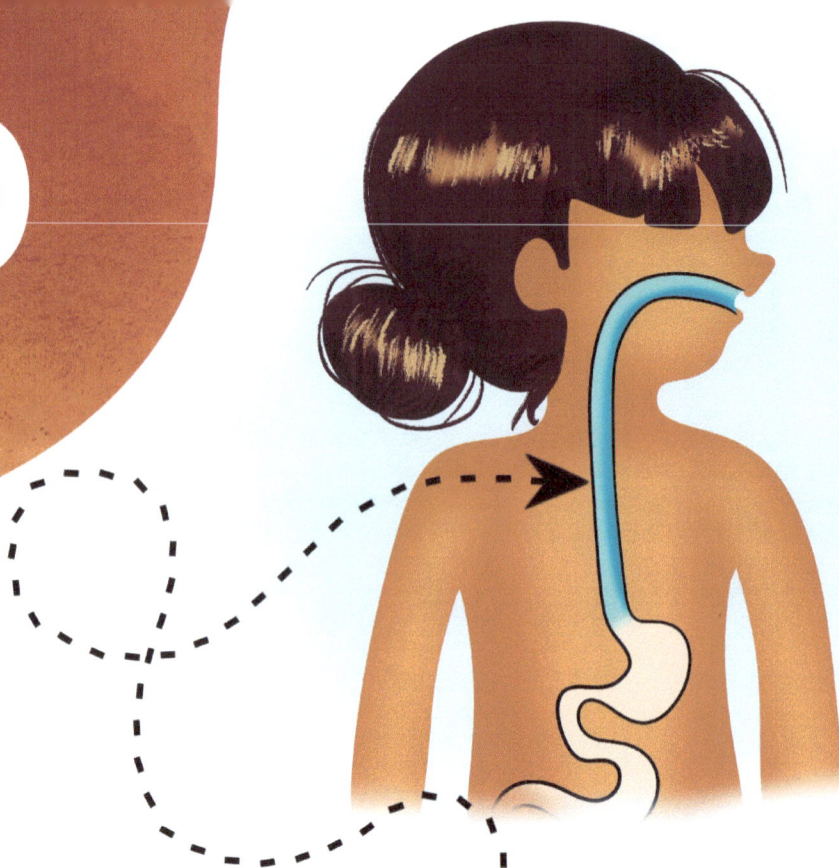

The **OESOPHAGUS** is the tube
where food moves down,
to your stomach
from your mouth.

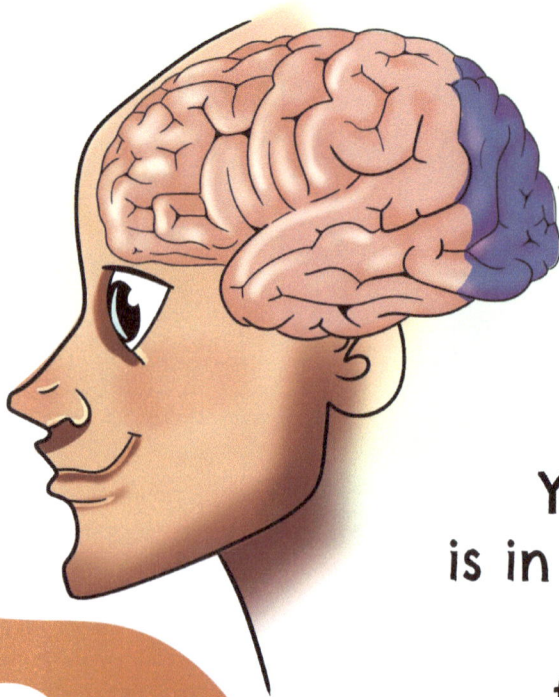

Your **OCCIPITAL LOBE**
is in the back of your brain.
You need it to see
from north to south.

O

O

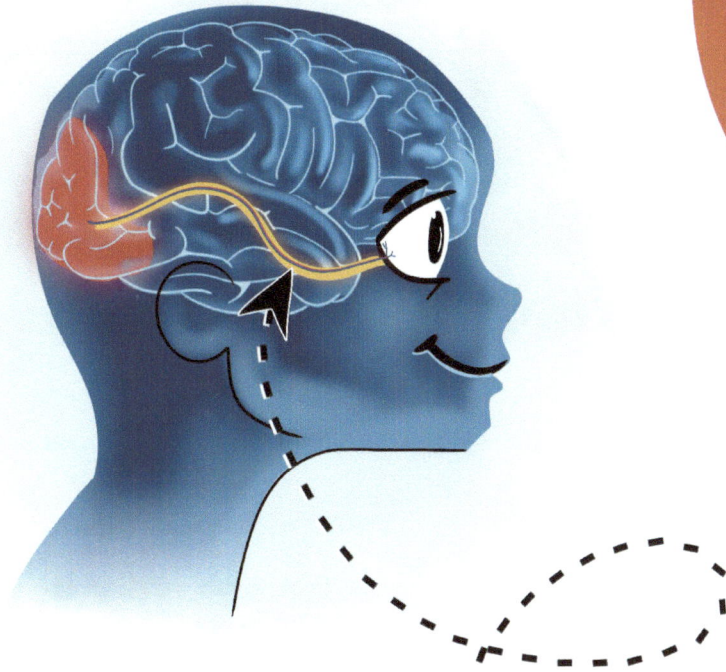

The **OPTIC NERVE**
transmits important information
from your eyes
and to your brain.

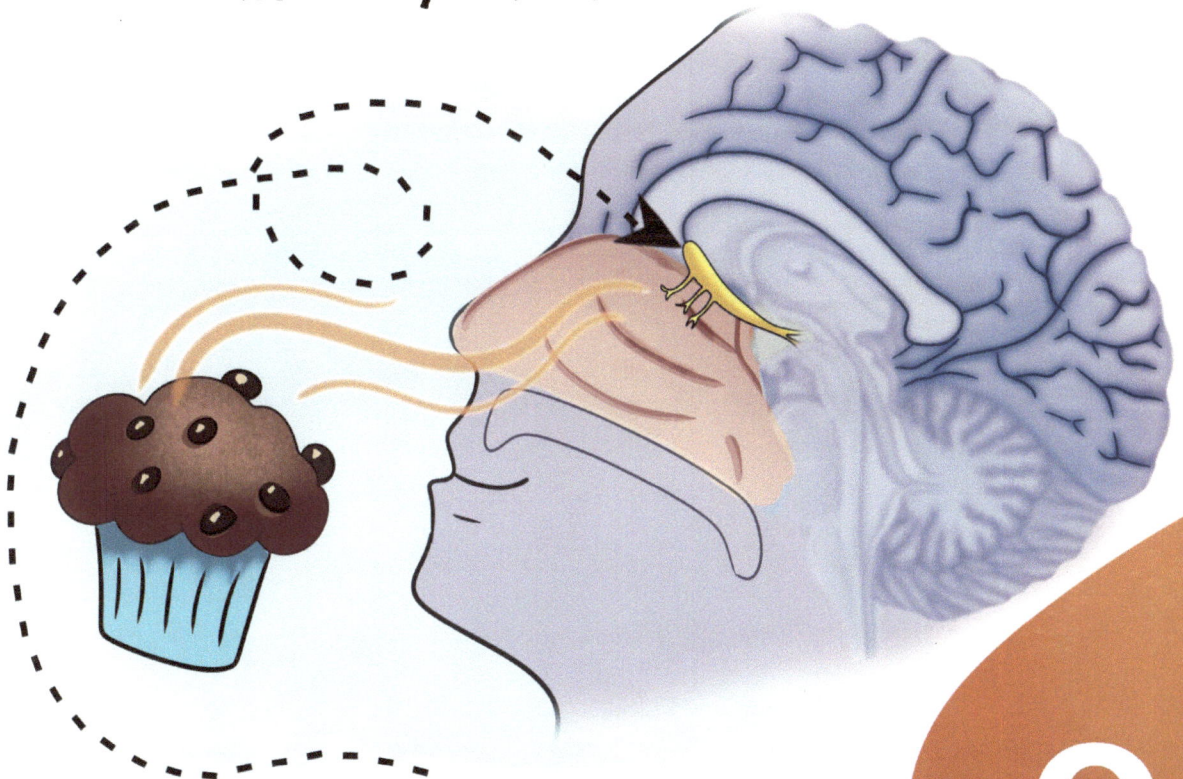

Your **OLFACTORY NERVE**
allows you to smell and taste,
all the food when you entertain!

O

P

Anterior   Posterior

POSTERIOR is the back
of everything,
from your head
down to your feet.

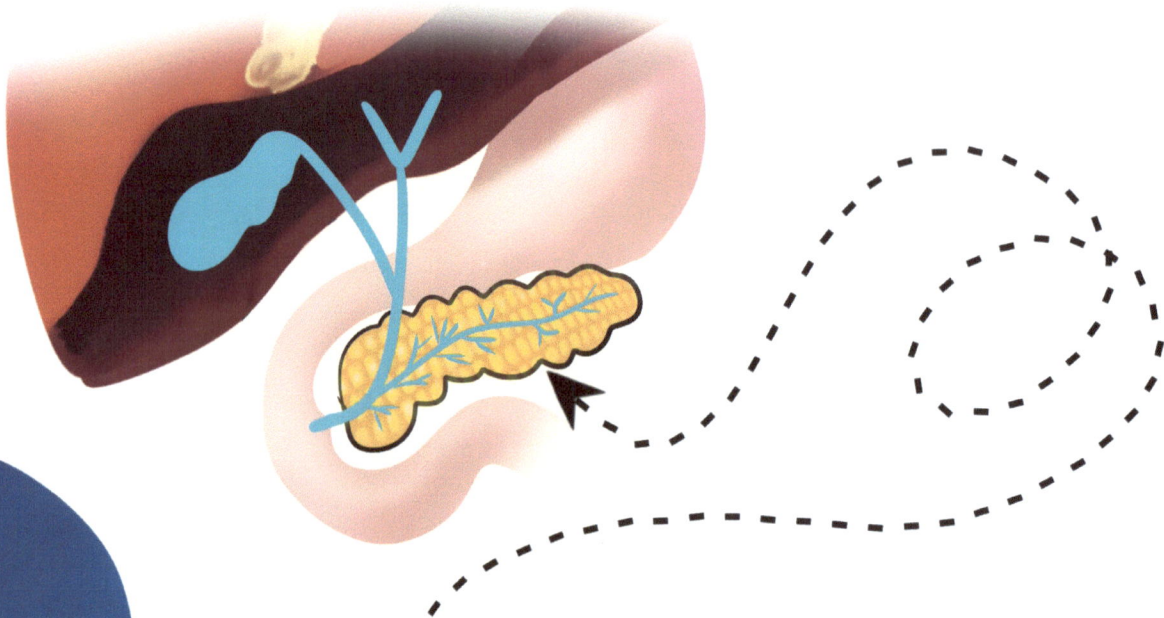

The **PANCREAS** in the belly produces insulin,
which is needed
when you eat something sweet.

p

P

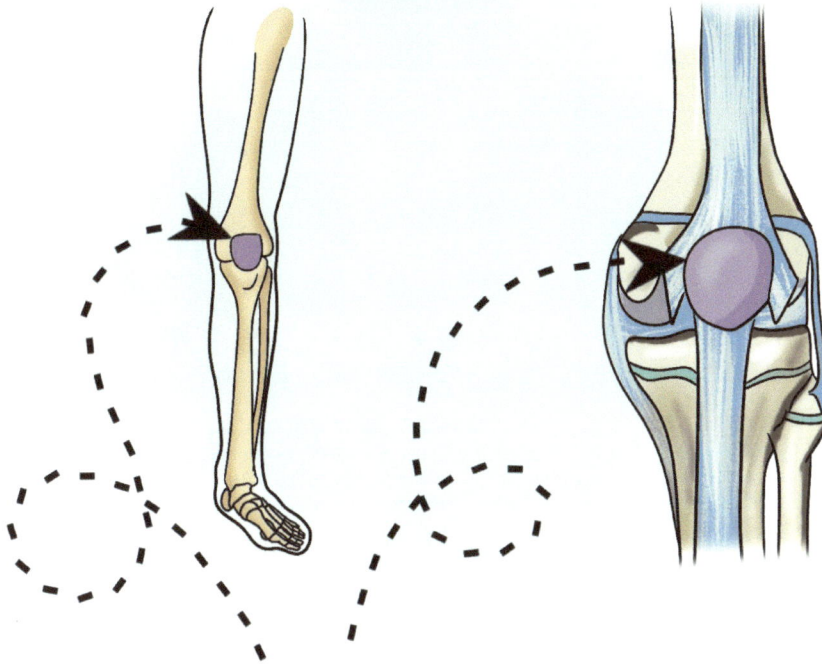

The **PATELLA** is your kneecap,
that glides
up and down when you move your knee.

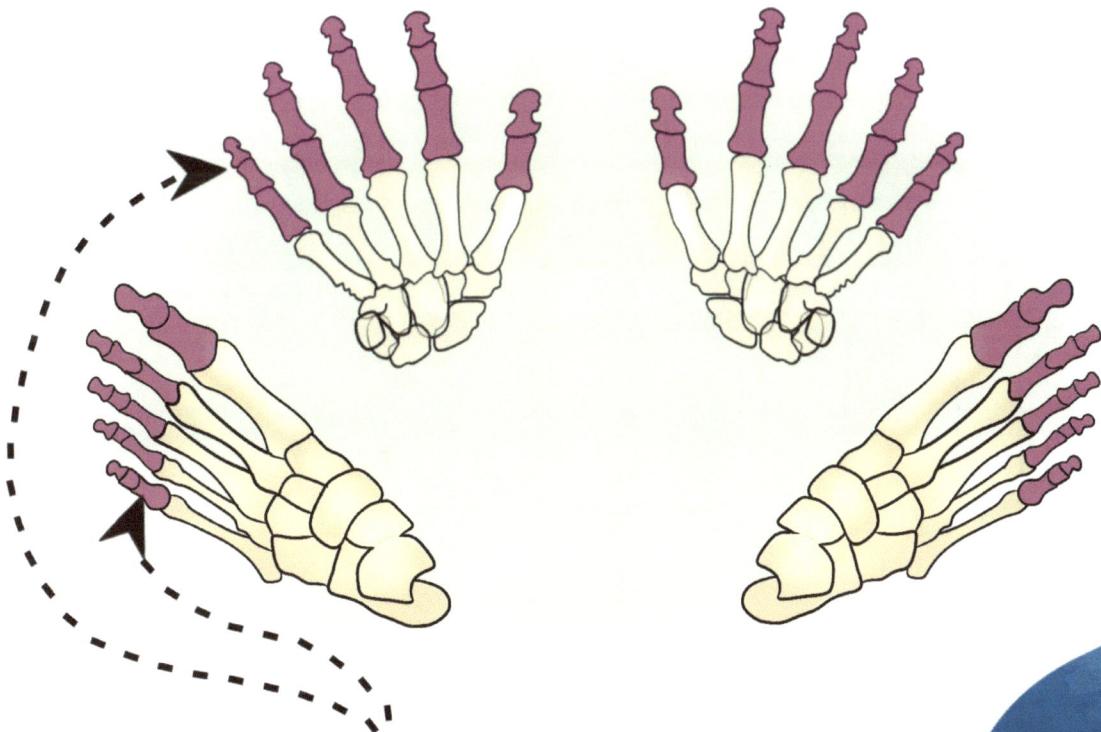

PHALANGES are in
your fingers and toes.
You have 56 altogether.
Count them and see!

p

Q

q

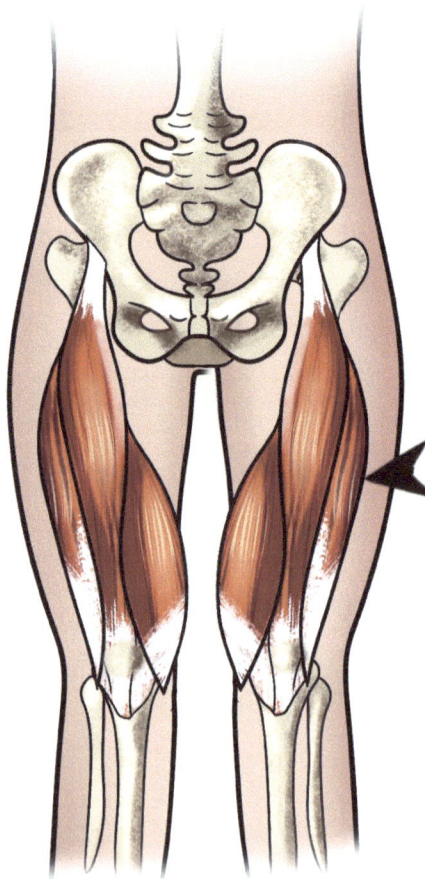

Your **QUADRICEPS** muscles
are at the front of each thigh.
They help you to stand up
and then sit down.

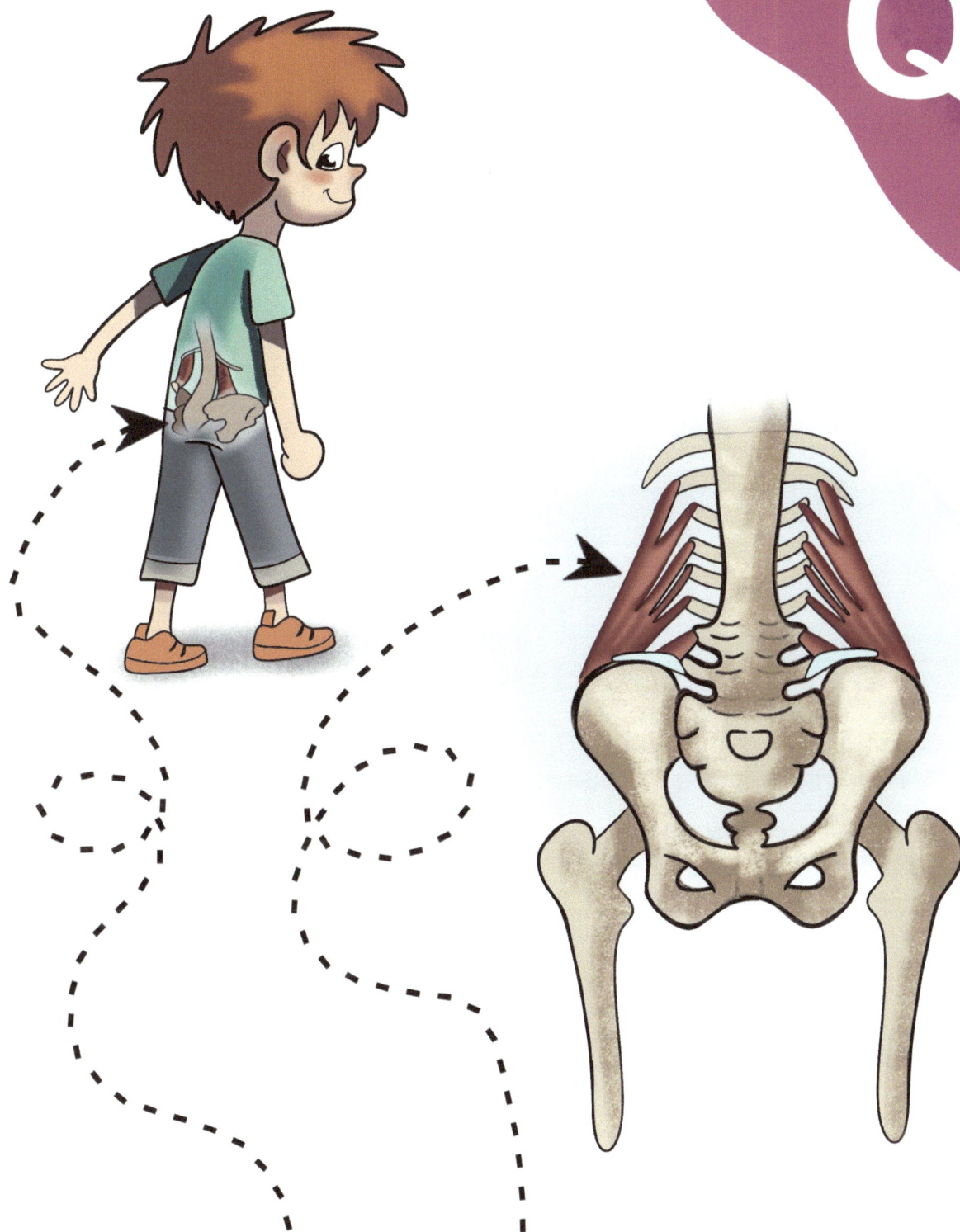

The **QUADRATUS LUMBORUM** muscle
keeps your pelvis and back straight,
whether you are still
or walking around.

R

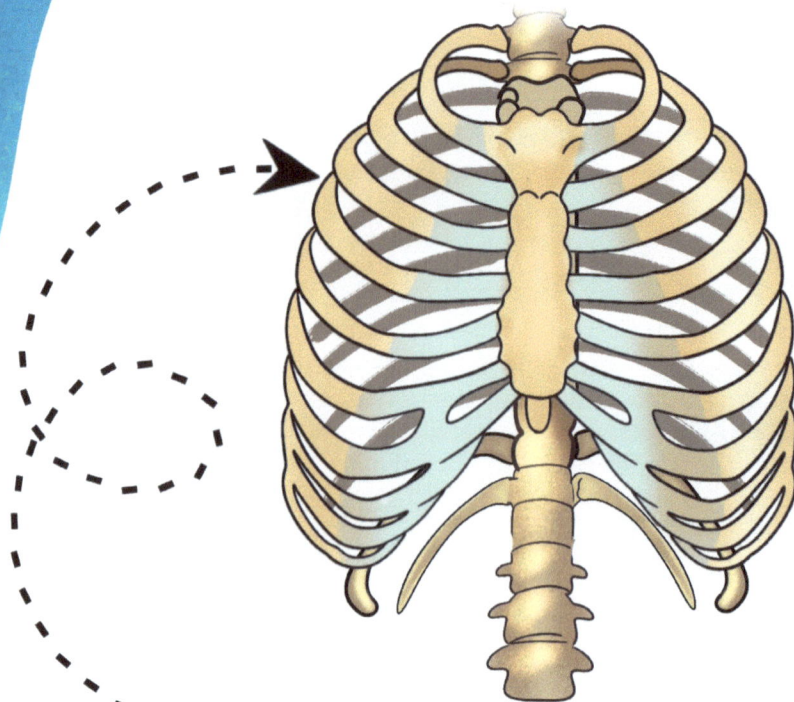

Your **RIBS** protect
your heart and your lungs.
There are 12 on the left
and 12 on the right.

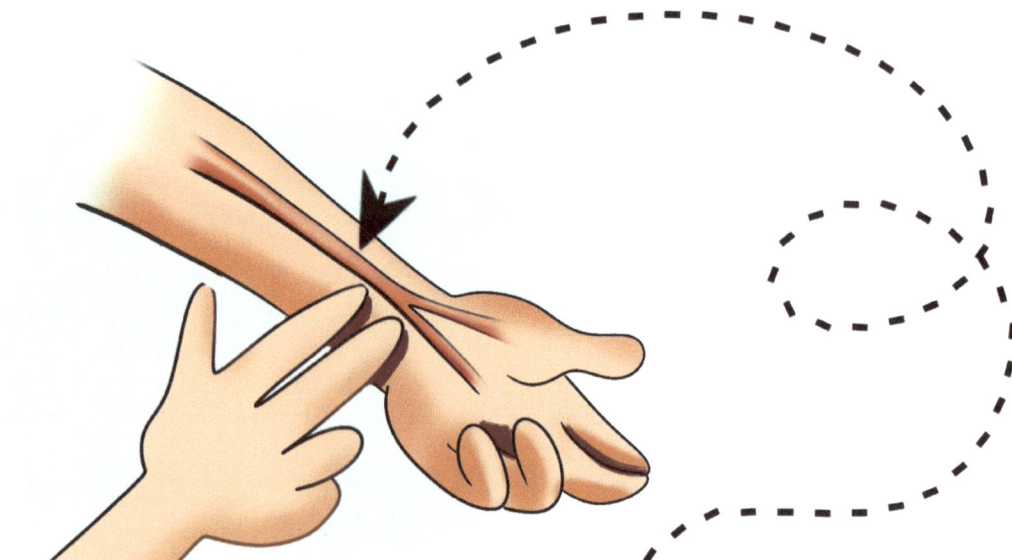

The **RADIAL ARTERY**
passes through your wrist.
It's where you feel your pulse
to check you are alright.

r

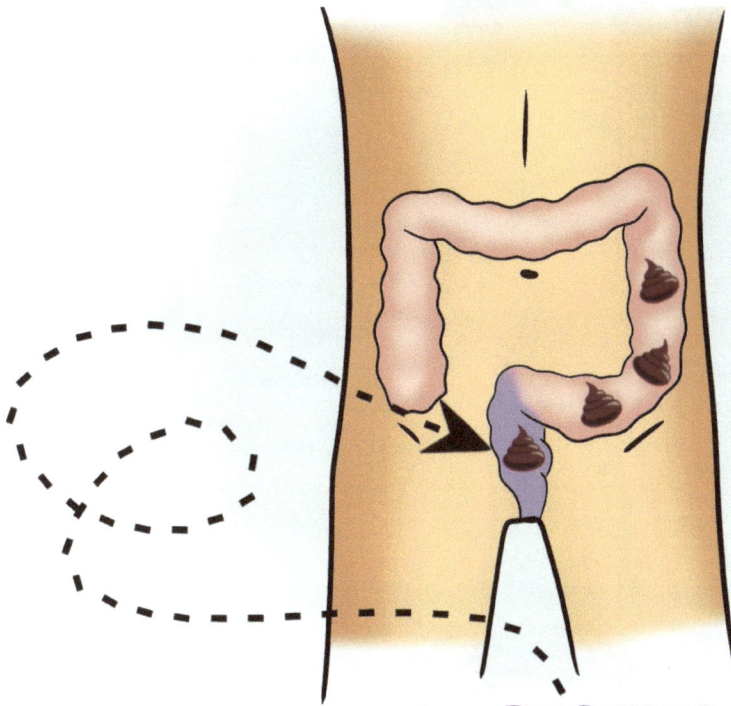

The **RECTUM** is
the end part of the bowel.
It is there
where your faeces is stored.

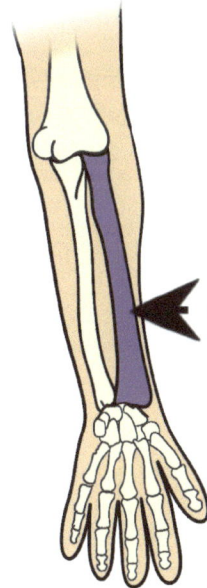

Your **RADIUS** is a bone
that is in your forearm.
It can be injured
if you fall off your skateboard!

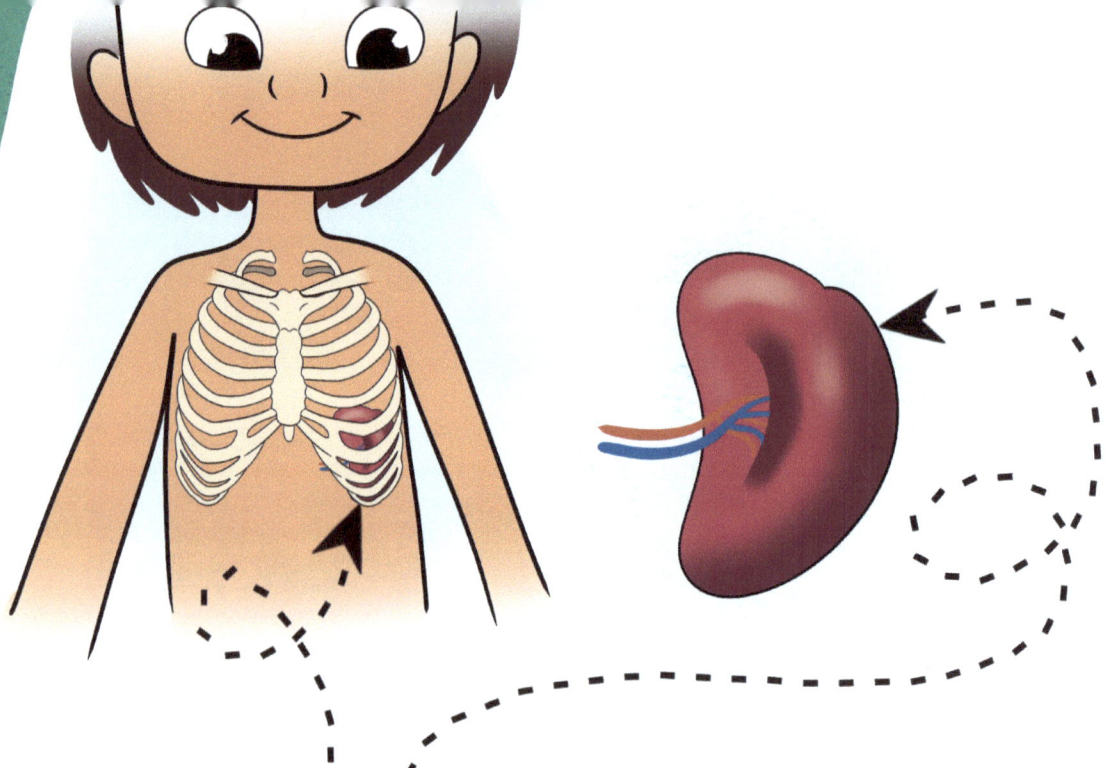

Your **SPLEEN** is in your belly.
It removes old red blood cells
and helps you
to fight infection.

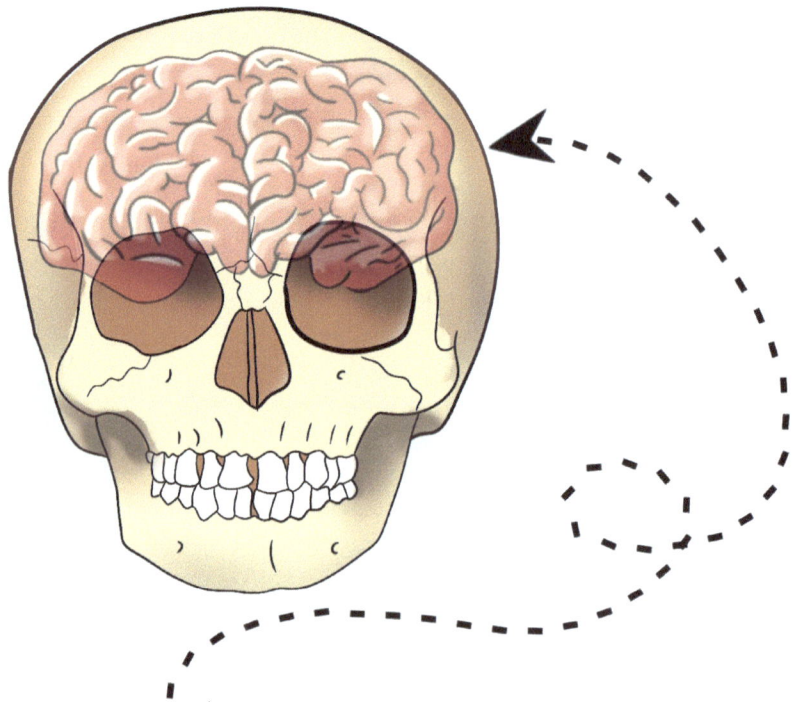

Your **SKULL** is a strong bone
that gives your head shape
and gives your fragile brain protection.

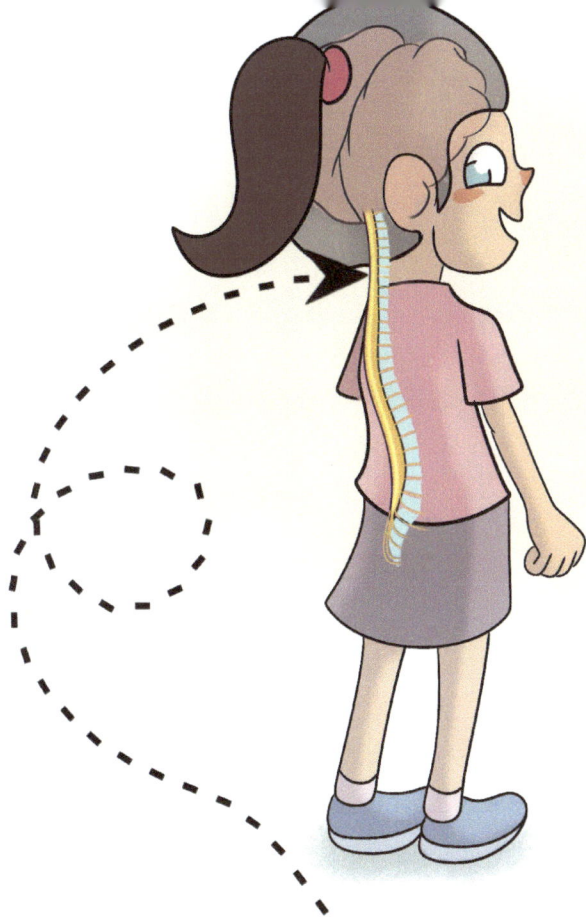

Your **SPINAL CORD**
runs down to your lower back,
from all the way up in your neck.

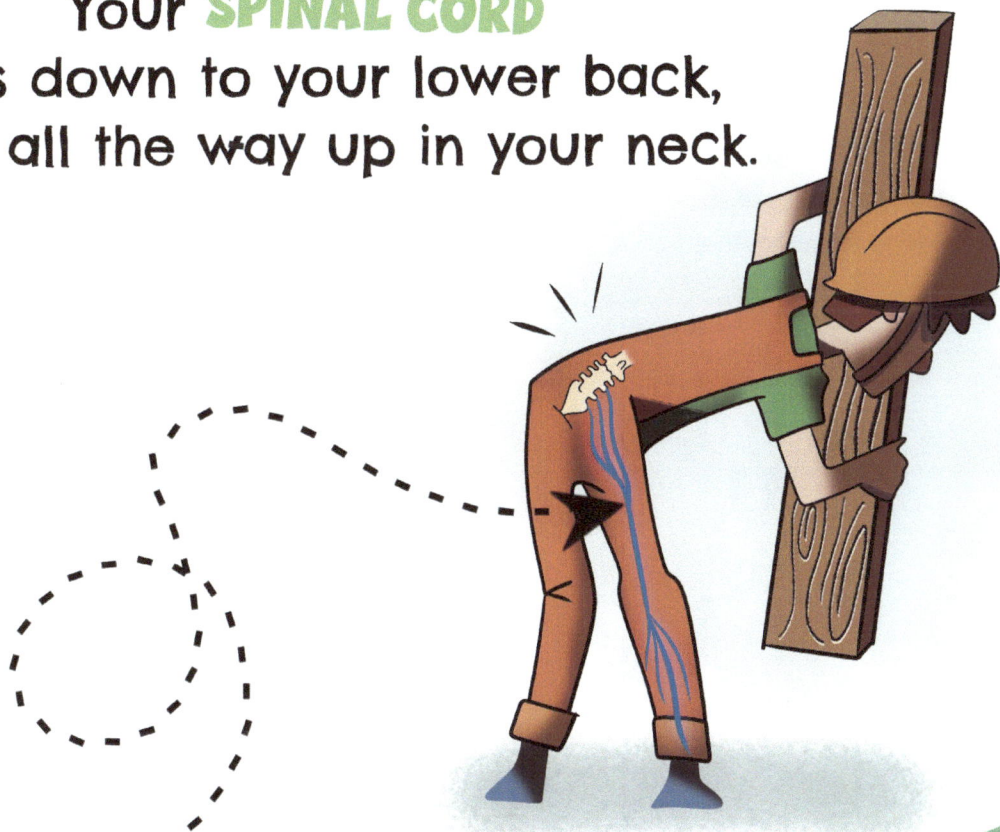

The **SCIATIC NERVE** can be injured
if you hurt your back
while working on the deck.

T

t

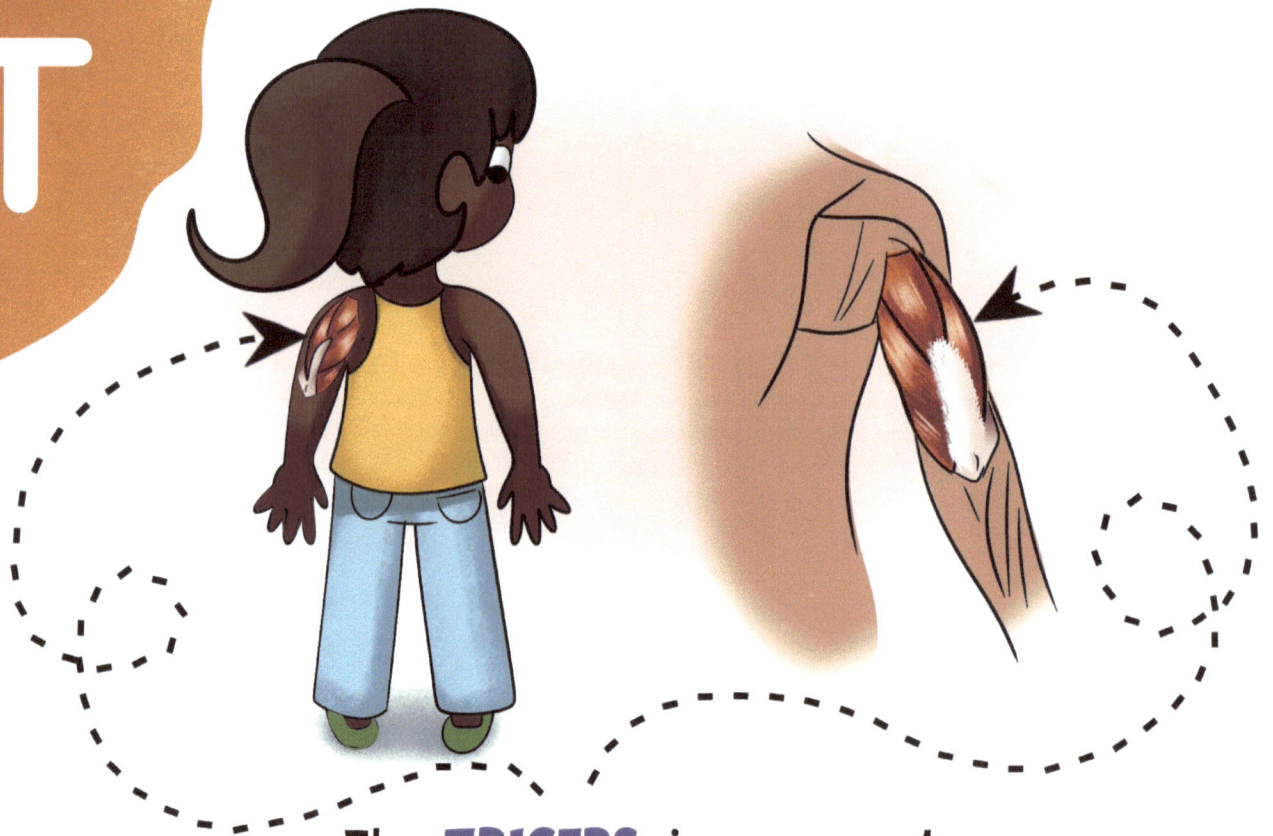

The **TRICEPS** is a muscle that helps you straighten your arm when it is bent.

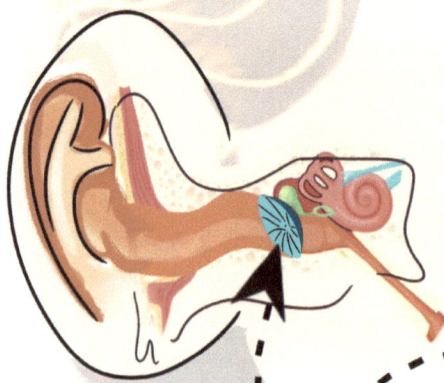

Your **TYMPANIC MEMBRANE** is in your ear.
It vibrates when sound waves from the air are sent.

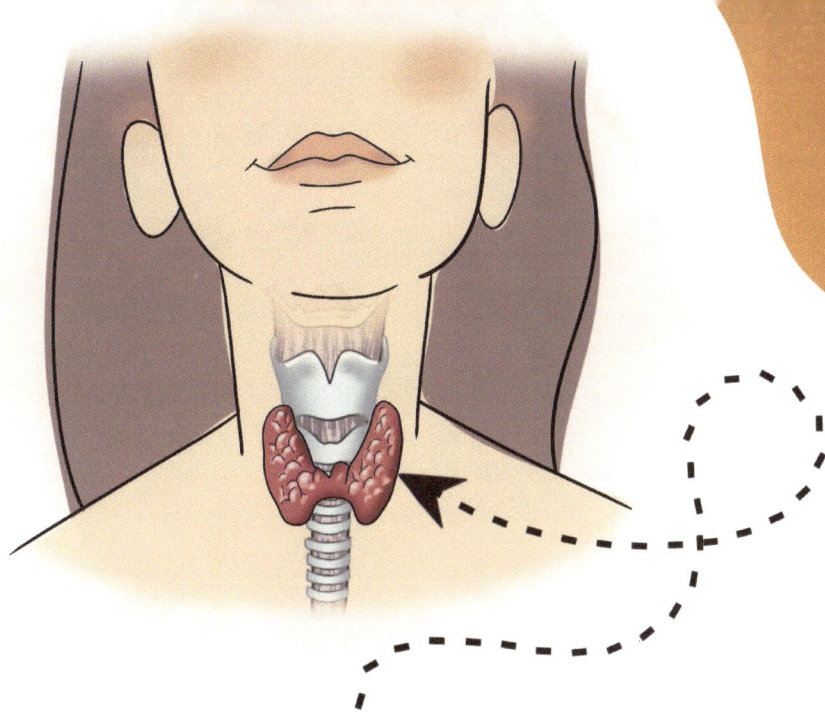

Your **THYROID** gland
is in your neck.
It's in the shape
of a butterfly.

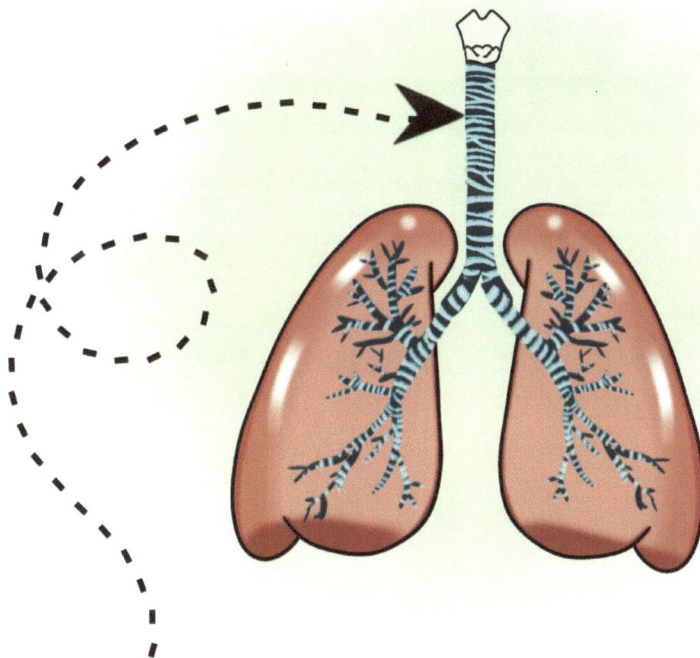

The **TRACHEA** is the windpipe
where air goes,
to your lungs even when you sigh.

u

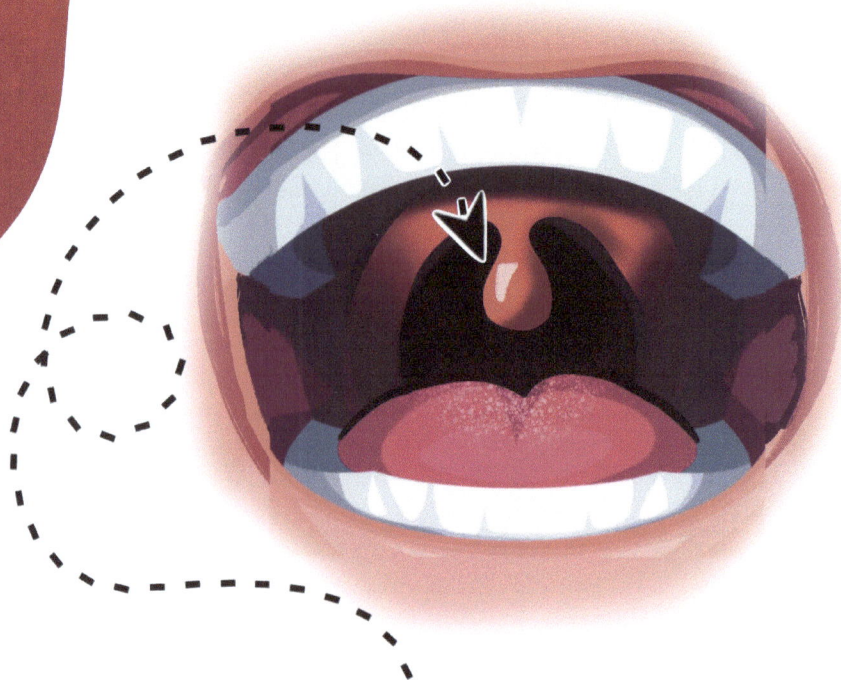

The **UVULA** is
the fleshy little ball.
It hangs
at the back of your throat.

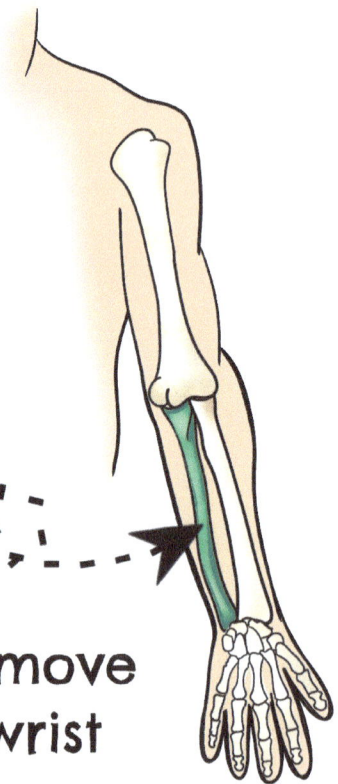

Your **ULNA** helps you move
at your elbow and wrist
and do things with your arms
like row a boat!

u

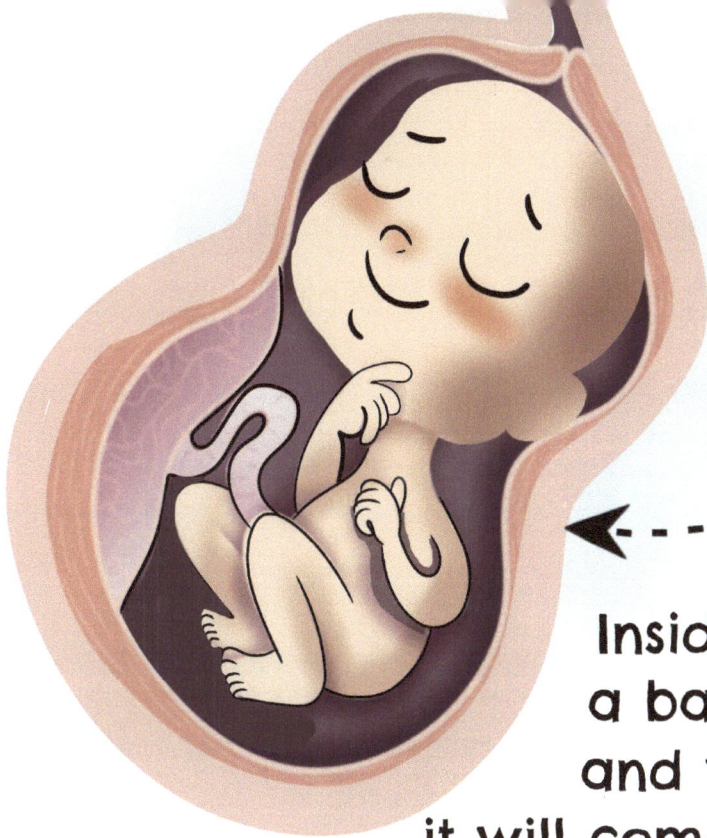

Inside the **UTERUS**
a baby can grow,
and when it's time
it will come out soft and cute.

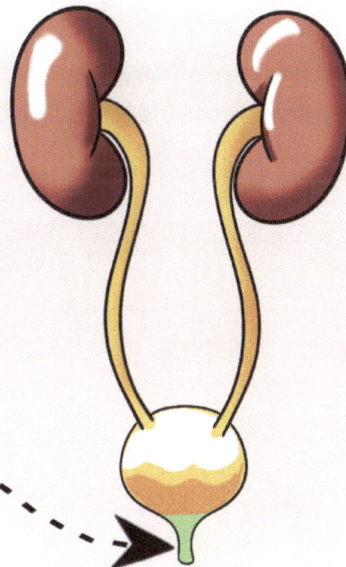

Urine from the bladder
passes through the **URETHRA**.
Drinking lots of water
will make your urine dilute.

V

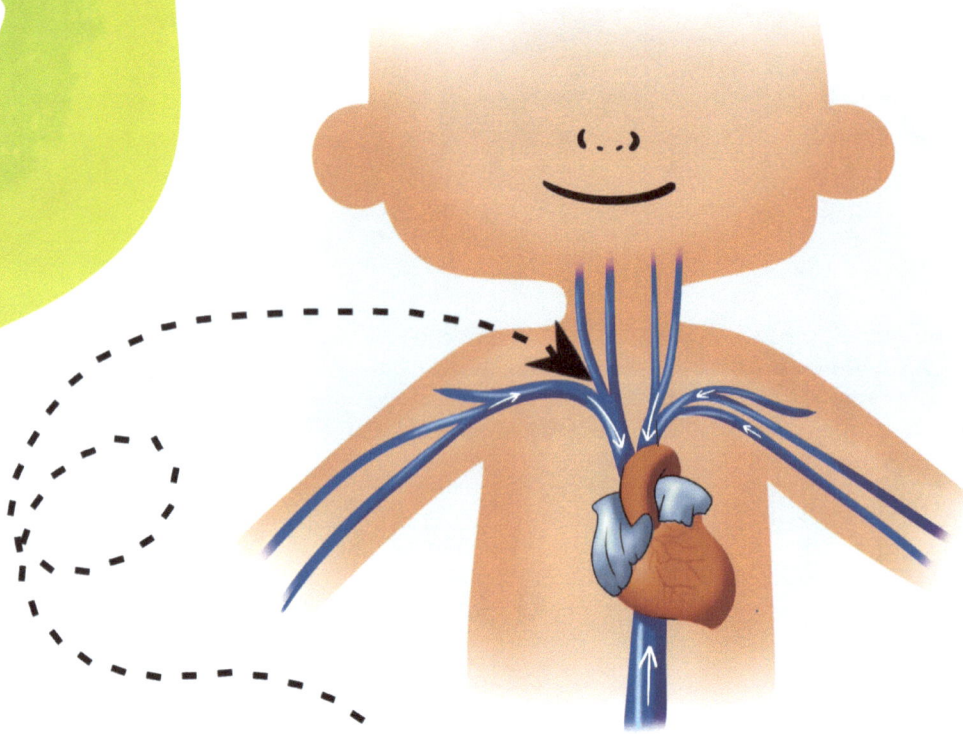

You have **VEINS** all over your body.
From your organs to your heart
is the direction of blood flow.

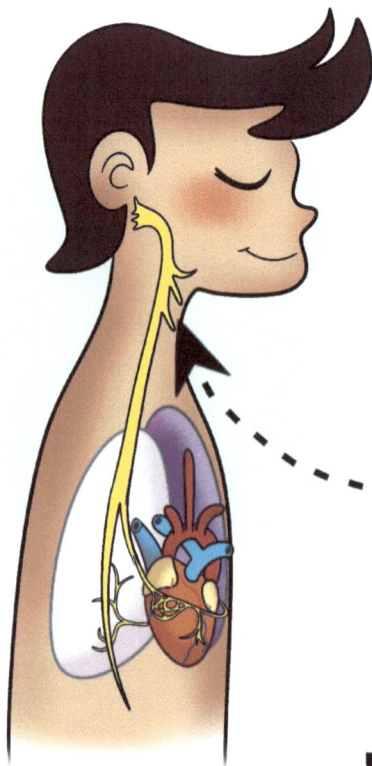

The **VAGUS NERVE**
from the brainstem
makes you feel relaxed
by keeping your breaths
and heart rate low.

v

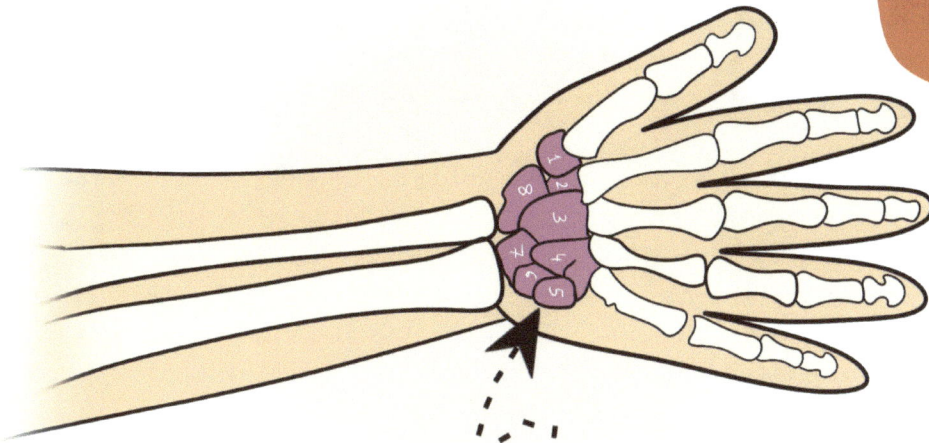

Your **WRIST** is made up
of small carpal bones.
There are eight to be exact.

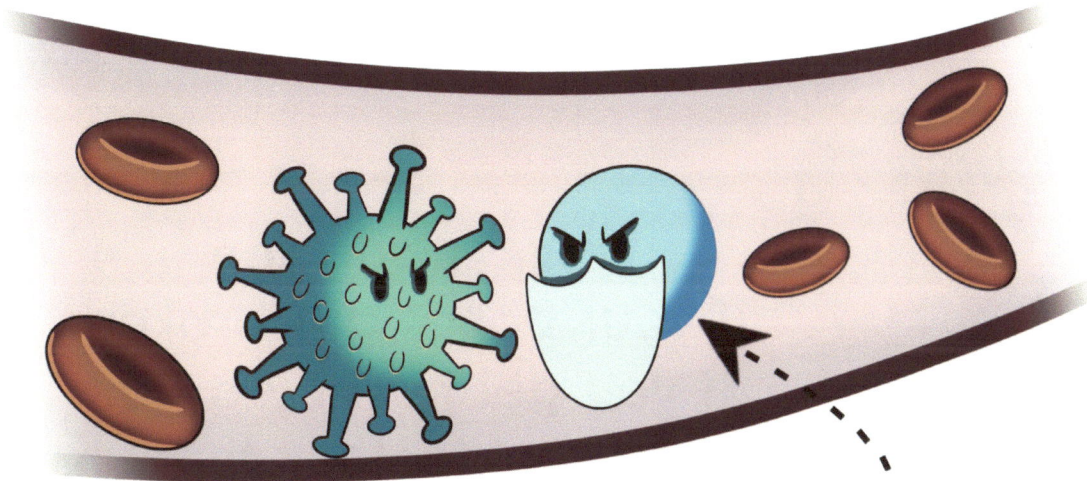

Your **WHITE BLOOD CELLS**
help make you better.
If you have an infection
they will quickly react!

X

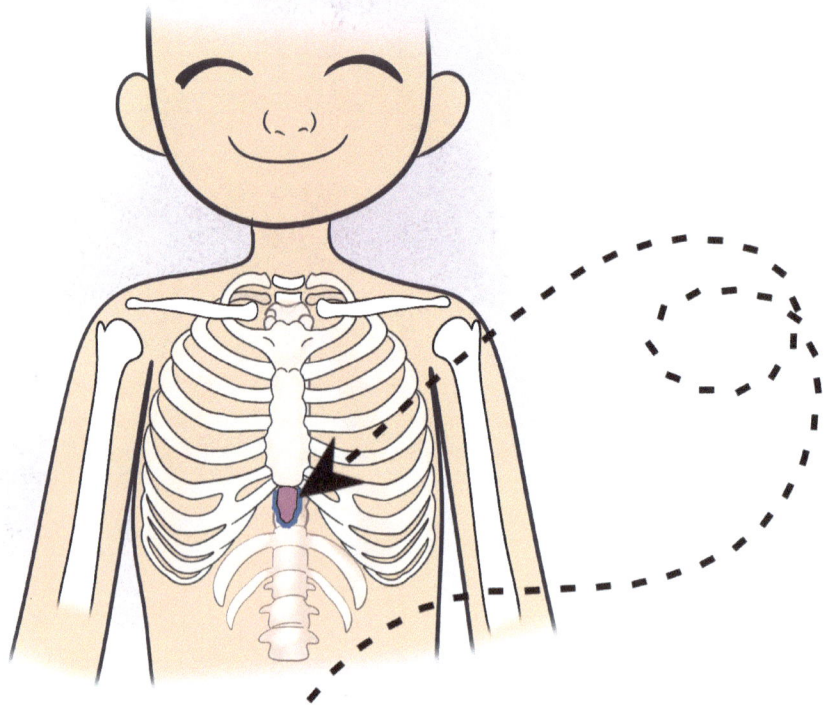

Your **XIPHOID** PROCESS
is a bony tip
that you can feel
at the end of your breastbone.

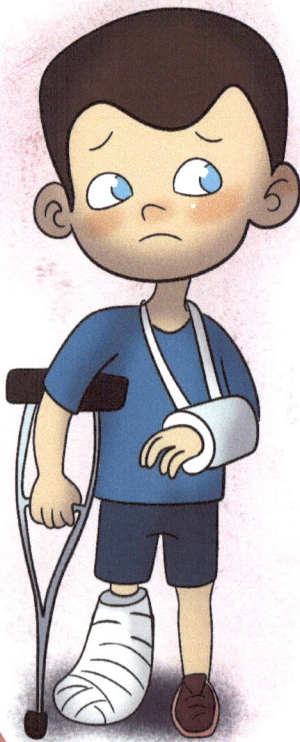

X-RAYS can see
if your bones are broken,
if you get hurt by a ball
that's accidentally thrown.

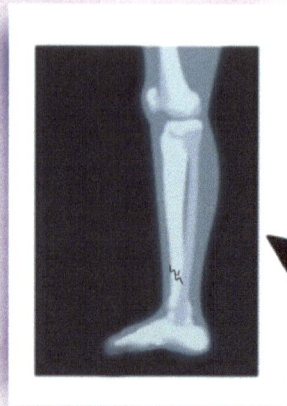

X

Y

You will see a **YOLK SAC** in a mother's womb when she is pregnant with a bundle of joy.

X Y

If you have a **Y CHROMOSOME** along with an X chromosome, when you are born you will look like a boy.

y

Z

The **ZYGOMATIC ARCH** is a bone that forms your cheekbone at the side, towards your ear.

Z

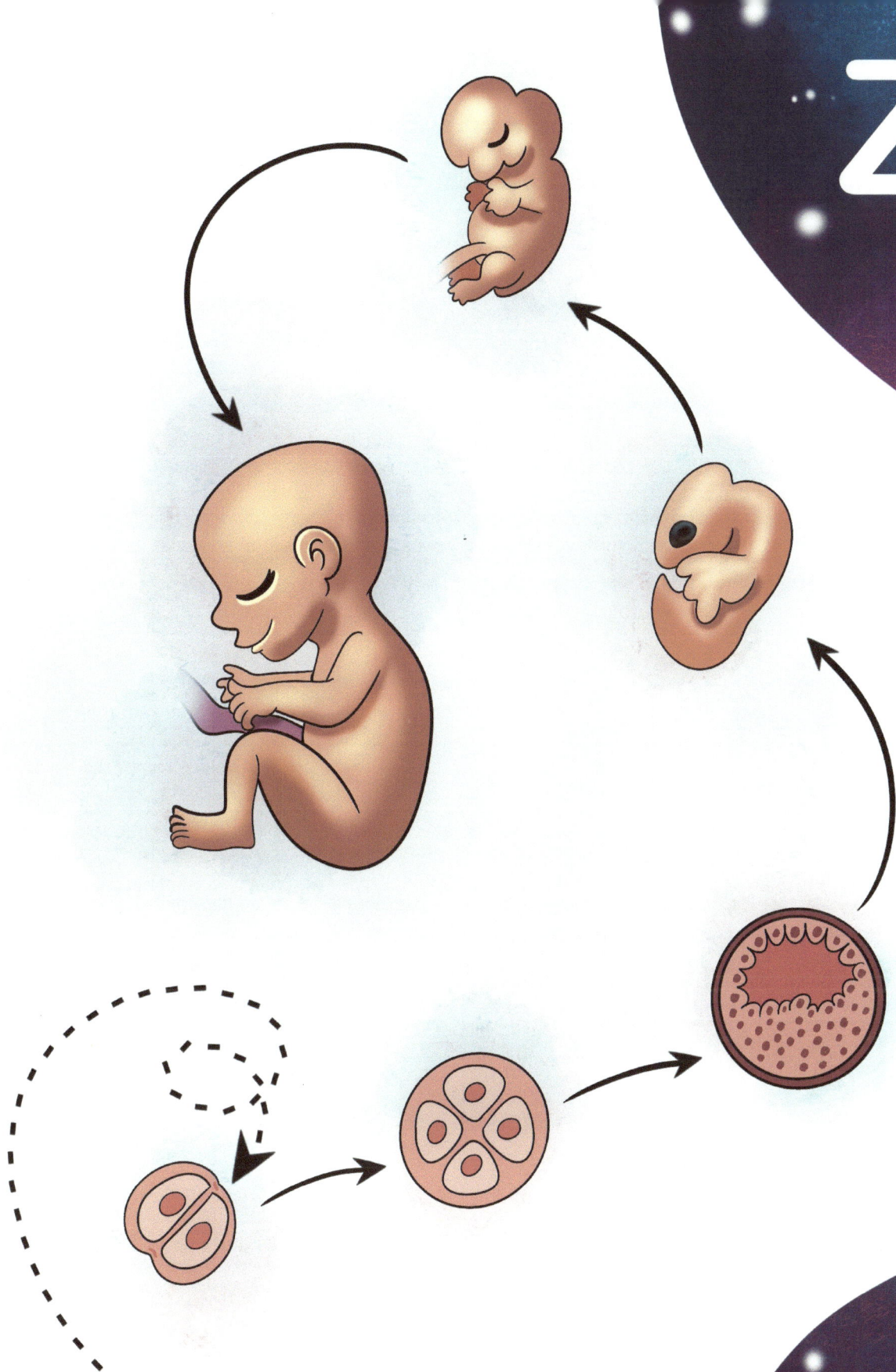

A **ZYCOTE** is a tiny fertilised egg
that grows into a baby
in less than year!

Z

z

www.ingramcontent.com/pod-product-compliance
Lightning Source LLC
Chambersburg PA
CBHW060847270326
41934CB00002B/36